III International Conference of Young Scientists on Contemporary Problems of Materials and Constructions (ICCPMC 2019)

IOP Conference Series: Materials Science and Engineering
Volume 684

Ulan-Ude, Russia
24 - 28 August 2019

ISBN: 978-1-7138-0936-4
ISSN: 1757-8981

Printed from e-media with permission by:

Curran Associates, Inc.
57 Morehouse Lane
Red Hook, NY 12571

Some format issues inherent in the e-media version may also appear in this print version.

Institute of Physics
Dirac House, Temple Back
Bristol BS1 6BE UK

Phone: 44 1 17 929 7481
Fax: 44 1 17 920 0979

techtracking@iop.org

Additional copies of this publication are available from:

Curran Associates, Inc.
57 Morehouse Lane
Red Hook, NY 12571 USA
Phone: 845-758-0400
Fax: 845-758-2633
Email: curran@proceedings.com
Web: www.proceedings.com

TABLE OF CONTENTS

IOP Conf. Series: Materials Science and Engineering **684** (2019) 011001 doi:10.1088/1757-899X/684/1/011001

Preface

In August 24-28, 2019 East Siberia State University of Technology and Management held the III International Conference of Young Scientists on Contemporary Problems of Materials and Constructions. The primary goal of the conference is to present new research results in such fields of study as functional and structural materials, mechanics of composite materials and constructions, computer modeling of materials and constructions, carried out by Russian and international scientific and educational institutions. The five-day conference brought together the specialists from different cities and organizations and provided an excellent opportunity to exchange knowledge, make oral contributions and poster presentations and to initiate discussions on the topics that are of interest to the conference participants. The conference was attended by Prof. Dr. Yurii Matvienko (head of Strength, Survivability and Safety of Machines Department of Mechanical Engineering Research Institute of the Russian Academy of Sciences), Assist. Prof. Dr. Oleg Mitrofanov (deputy head of the Research Department of Strength – deputy chief designer for strength of JSC "Sukhoi Civil Aircraft"), Prof. Dr. Lyubov Bokhoeva (head of Strength of Materials Department of East Siberia State University of Technology and Management) and other famous scientists. The scientists showed their great potential in studying various problems in fields of mechanics of materials and constructions, fracture mechanics, strength, viability and safety of machines, materials science as well as applying new approaches, which develop and improve methods of mechanics.

The Conference also provides an interdisciplinary round table for researchers, practitioners, students and young scientists to present and discuss the most recent innovations and developments in the field of mechanical engineering, materials science, technology of machine-building productions.

The papers of the conference cover a wide range of technical areas and modern aspects of the mechanical engineering, computer modeling of materials and constructions, fracture mechanics, material science and new technologies in industry.

The organizing committee is grateful to the hotel "Baikal Riviera" (Russia, republic of Buryatia, Pribaykalsky district, Gremyachinsk village), on the basis of which the conference was held. The committee is thankful to the employees of Buryat State University having made an outstanding contribution to organizing and holding the conference. The organizing committee is expressing sincere gratitude to Prof. Dr. Aleksei Mikhailovich Pokrovskii (BMSTU), Dr. Igor Petrovich Aistov (OmSTU), Dr. Vitalii Evdokimovich Rogov (BINM SB RAS) for peer reviewing scientific articles for publication in the collection.

IOP Conf. Series: Materials Science and Engineering **684** (2019) 011002 doi:10.1088/1757-899X/684/1/011002

III International Conference of Young Scientists on Contemporary Problems of Materials and Constructions photographs

IOP Conf. Series: Materials Science and Engineering **684** (2019) 011002 doi:10.1088/1757-899X/684/1/011002

IOP Conf. Series: Materials Science and Engineering **684** (2019) 011002 doi:10.1088/1757-899X/684/1/011002

IOP Conf. Series: Materials Science and Engineering **684** (2019) 011002 doi:10.1088/1757-899X/684/1/011002

IOP Conf. Series: Materials Science and Engineering **684** (2019) 011002 doi:10.1088/1757-899X/684/1/011002

IOP Conf. Series: Materials Science and Engineering **684** (2019) 011002 doi:10.1088/1757-899X/684/1/011002

IOP Conf. Series: Materials Science and Engineering **684** (2019) 011002 doi:10.1088/1757-899X/684/1/011002

IOP Conf. Series: Materials Science and Engineering **684** (2019) 011002 doi:10.1088/1757-899X/684/1/011002

IOP Conf. Series: Materials Science and Engineering **684** (2019) 011002 doi:10.1088/1757-899X/684/1/011002

IOP Conf. Series: Materials Science and Engineering **684** (2019) 011002 doi:10.1088/1757-899X/684/1/011002

Comparison of J-A field and FEM results
HRR field (J-integral only) considerably deviates from FEM results
J-A closely follows finite element results

IOP Conf. Series: Materials Science and Engineering **684** (2019) 011002 doi:10.1088/1757-899X/684/1/011002

IOP Conf. Series: Materials Science and Engineering **684** (2019) 011002 doi:10.1088/1757-899X/684/1/011002

IOP Conf. Series: Materials Science and Engineering **684** (2019) 011002 doi:10.1088/1757-899X/684/1/011002

IOP Conf. Series: Materials Science and Engineering **684** (2019) 011003 doi:10.1088/1757-899X/684/1/011003

Funding acknowledgements

The III International Conference of Young Scientists on Contemporary Problems of Materials and Constructions was held with a grant from the Russian Foundation for Basic Research; project No. 19-08-20052; contract No. 19-08-20052\19.

IOP Conf. Series: Materials Science and Engineering **684** (2019) 011004 doi:10.1088/1757-899X/684/1/011004

List of organizing committee members of III International Conference of Young Scientists on Contemporary Problems of Materials and Constructions (ICCPMC)

- **Chairman:** Prof. Dr. ***I.G. Sizov***, East Siberia State University of Technology and Management, Ulan-Ude, Russia
- **Co-chair:** Prof. Dr. ***L.A. Bokhoeva,*** East Siberia State University of Technology and Management, Ulan-Ude, Russia
- **Deputy Chairman:** Assist. Prof. Cand. ***E.B. Bochektueva,*** East Siberia State University of Technology and Management, Ulan-Ude, Russia

- Assist. Prof. Cand. ***S.A. Bazaron,*** East Siberia State University of Technology and Management, Ulan-Ude, Russia
- Cand. ***A.S. Chermoshentseva,*** Bauman Moscow State Technical University, Moscow, Russia
- Sen. Lect. ***N.N. Anchiloev,*** East Siberia State University of Technology and Management, Ulan-Ude, Russia
- Sen. Lect. ***A.B. Baldanov,*** East Siberia State University of Technology and Management, Ulan-Ude, Russia
- Sen. Lect. ***V.Yu. Kurokhtin,*** East Siberia State University of Technology and Management, Ulan-Ude, Russia
- Train. Mast. ***Ya.A. Shevchenko,*** East Siberia State University of Technology and Management, Ulan-Ude, Russia

IOP Conf. Series: Materials Science and Engineering **684** (2019) 011005 doi:10.1088/1757-899X/684/1/011005

Logotype of III International Conference of Young Scientists on Contemporary Problems of Materials and Constructions (ICCPMC)

IOP Conf. Series: Materials Science and Engineering **684** (2019) 011006 doi:10.1088/1757-899X/684/1/011006

Peer review statement

All papers published in this volume of *IOP Conference Series: Materials Science and Engineering* have been peer reviewed through processes administered by the proceedings Editors. Reviews were conducted by expert referees to the professional and scientific standards expected of a proceedings journal published by IOP Publishing.

Formation of the properties of nano-modified concretes

A V Sharanova, M A Dmitrieva and A A Zakharov

Immanuel Kant Baltic Federal University, 14 Nevskiy St, Kaliningrad 236016, Russia

E-mail: a.sharanova@gmail.com

Abstract. Nanomodified concrete is obtained from multicomponent dry building mixes based on Portland cement containing nanoscale filler particles. This paper examines the cement-sand mixture activated in a high-energy mill as a nanomodifier. The result is the gain in strength of concrete in the initial hardening time. The influence of fractional characteristics of sand on the activation process is analyzed. Three sand samples of different deposits of the Kaliningrad region characterized by different size and content of dust and clay particles were studied. The results of tests of samples of fine-grained concrete with activated and non-activated components for the initial compressive strength at the age of 0-90 minutes after the preparation of the mixture are presented. It is noted that the mechanical activation of cement-sand mixture is most effective when there are large fractions in the activated sand.

1. Introduction

Modern construction composite materials are multicomponent and polydisperse systems that include various modifying additives, ultrafine components, etc. One of the key factors of nanotechnology in construction material science should be to achieve the desired or maximum quality of material at minimum cost resource. As a primary nanomaterial in this paper we consider a fine-ground mixture of Portland cement and sand, obtained by joint mechanical activation in a high-speed ball mill. Mechanical activation provides an increase in the speed of concrete setting and more complete use of cement binding properties [1]. According to the results of previous works, it was proved that this type of activation is the most effective [2]. The concrete mixture containing fine ground cement-sand mixture sets faster and shows a significant increase in strength in the initial setting period. This mixture can be used in construction 3D printing with concrete as it meets the requirements for strength to withstand the weight of the above-printed layers [3-6].

2. Materials and methods

Sands from various locations of the Kaliningrad region are used as the components for activation: sea sand (SS) of the medium size (Fineness modulus = 2.1); sand from the quarry near village Sinyavino (SQS) of the fine size (Fineness modulus = 2.0); sand from the quarry Rybachye (SQR) of the very fine size (Fineness modulus = 1.34). The size of sand was determined according to GOST 8735-88 through the fineness modulus. Conducted joint mechanical activation of cement and sand in the RetschEMax high speed ball mill measured initial strength of the samples. According to the results of earlier works on the selection of composition for additive technologies in construction, the optimal composition meeting the requirements of 3D printing with concrete was determined [7]. Mixtures of fine-grained concrete with activated and non-activated components are made of this composition.

2.1. Materials

The following materials were used for the preparation of samples:

- Portland cement Eurocem super 500, CEM I 42.5H, Petersburg Cement;
- Sand: for the composition No. 1: sea sand from the coast of the urban-type settlement Yantarny, Kaliningrad region; for the composition No. 2: sand from the quarry near village Sinyavino, Kaliningrad region; for the composition No. 3: sand from the quarry Rybachye, Kaliningrad region.
- Polypropylene fiber: length of 12 mm, thickness of 35 μm, country of origin: Russia;
- Silica fume (SF): wastes of the metallurgical industry, country of origin: Poland;
- Highly active metakaolin (MK): white, country of origin: Russia, Chelyabinsk region, Synergo;
- STACHEMENT 1267 hyperplasticizer based on polycarboxylates for the production of ready-mixed concrete: country of origin: Poland.

The consumption of materials per 1 m^3 of the concrete mixture is given in table 1.

Table 1. Materials per 1 m^3.

W/C ratio	Cement (kg m^{-3})	Sand (kg m^{-3})	Water (l m^{-3})	Silica fume (SF) (kg m^{-3})	Metakaolin (MK) (kg m^{-3})	Fiber (kg m^{-3})	Hyperplast. (l m^{-3})
0.66	545	1168	360	156	78	1.2	11.7

Characteristics of sand are presented in table 2. All the characteristics were determined according to Russian State standards GOST 8736-2014 "Sand for construction work. Specifications" [8] and GOST 8735-88 "Sand for construction work. Testing methods" [9].

Table 2. Sand characteristics.

Location of origin	Symbol	Size group	Fineness modulus	Content of dust and clay particles (%)	True density (kg m^{-3})	Bulk density, kg m^{-3}	Voidness, %
Coast	SS	Medium	2.1	-	2612	1474	43.6
Quarry Sinyavino	SQS	Fine	2	2.2	2642.5	1330	49.7
Quarry Rybachye	SQR	Very fine	1.34	3.6	2551	1501	41.2

The images of the sand are shown in figures 1-3. The images were obtained using an optical electron microscope.

Figure 1. Sea sand (SS).

Figure 2. Sinyavino quarry sand (SQS).

Figure 3. Rybachye quarry sand (SQR).

2.2. Methods

Joint mechanical activation of cement and sand was carried out in the Retsch EMax high-speed ball mill at a speed of 1000 rpm for 5 minutes, since this activation mode is the most effective [1]. Corundum balls with diameters of 6 and 10 mm in a ratio of 1:1 by weight were used as grinding media. The ratio of grindable mixture/grinding media is 1:2.

Six compositions of the initial building mixtures were considered: No. 1 SS – on the basis of non-activated mixture using SS sand; No. 2 CSSA – on the basis of activated mixture using SS sand; No. 3 SQS – on the basis of non-activated mixture using SQS sand; No. 4 CSQSA – on the basis of activated mixture using SQS sand; No. 5 SQR – on the basis of non-activated mixture using SQR sand; No. 6 CSQRA – on the basis of activated mixture using SQR sand.

Next, cubic samples of size 20 × 20 × 20 mm were made with activated and non-activated components in accordance with the consumption of materials presented in Table 1. The samples were tested for compressive strength at the initial setting time (from 0 to 90 minutes every 15 minutes) on an Instron Electropuls E1000 testing machine.

3. Results

The results of the strength tests of the samples at the initial setting time are shown in the graph in figure 4. In the graph, solid lines indicate the increase in strength of samples with non-activated components, the strength of samples with activated components is indicated by a dash line.

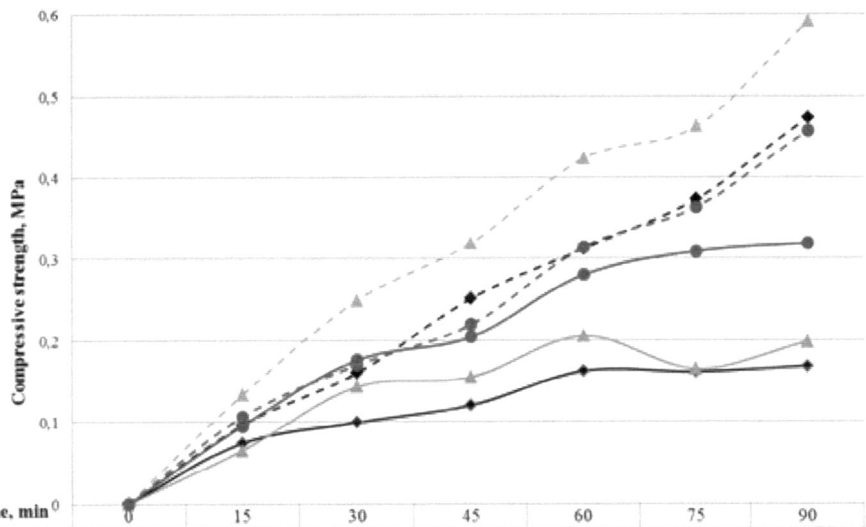

Setting time, min	0	15	30	45	60	75	90
No. 1 BS	0	0,074	0,099	0,12	0,162	0,161	0,168
No. 2 CSSA	0	0,096	0,16	0,251	0,312	0,373	0,473
No. 3 SQS	0	0,065	0,143	0,155	0,205	0,165	0,198
No. 4 CSSA	0	0,133	0,249	0,319	0,423	0,462	0,591
No. 5 RS	0	0,095	0,176	0,204	0,28	0,309	0,319
No. 6 CRSA	0	0,106	0,17	0,219	0,314	0,363	0,457

Figure 4. Compressive strength test results.

The dependence of the initial strength of samples with non-activated components (compositions No. 1, 3, 5) on the size of the sand is noted. Samples of composition No. 1 with medium-sized sea sand show the lowest strength in the first 90 minutes of hardening, samples of composition No. 5 with very fine sand from the Rybachye quarry have the greatest strength. Samples of composition No. 3 with fine sand show medium strength.

Compositions with joint activation of cement and sand (No. 2, 4, 6) show an increase in strength compared with the corresponding samples without activation (No. 1, 3, 5). There is a significant increase in strength in compositions No. 2 and No. 4. Strength at the age of 90 minutes for composition No. 2 increases by 2.81 times or by 181% relative to composition No. 1. The strength of composition No. 4 at the age of 90 minutes increases by 2.98 times or by 198.4% relative to composition No. 3. At the same time, there is no significant difference in strength between samples No. 5 and 6: at the age of 90 minutes, the strength of composition No. 6 increases by 1.43 times or by 43.3%. Therefore, the mechanical activation of the sands belonging to the group of very fine sand is ineffective. The test results show that the joint activation of cement and sand gives a significant increase in strength at the early setting time only when large fractions are present in the activated sand.

4. Conclusion

The nanomodified additives obtained by the joint activation of cement and sand in high-energy ball mills are considered. The use of such nanomodified additives leads to an increase in the strength of concrete in the initial periods of hardening. Analysis of the effect of the initial characteristics of sand on the activation process showed that the mechanical activation of the cement-sand mixture is most effective when the activated sand contains large fractions.

Acknowledgments

The reported study was funded by RFBR and Kaliningrad region, project number 19-48-390001 "Multicomponent nanomodifiers of concrete mixes".

References

[1] Andreeva A, Davidova N and Burenina O 2014 Joint activation of the mineral part of fine concrete *New materials and technologies in the Arctic (Yakutsk)* (Stavropol: Logos) pp 222–5

[2] Sharanova A, Lenkova D and Panfilova A 2018 Effect of mechanical activation of initial components on mechanical properties of concrete mixture *Aktualnye problemy sovremennogo stroitelstva: materialy 71 Vserossiiskoy nauchno-prakticheskoi konferencii studentov, aspirantov I molodikh uchenikh SPbGASU (Saint-Petersburg)* vol 2 (Saint-Petersburg: SPbGASU Press) pp 108–14

[3] Buswell R, Leal de Silva W, Jones S and Dirrenberger J 2018 3D printing using concrete extrusion: A roadmap for research *Cement and Concrete Research* **112** 37–49

[4] Panda B, Unluer C and Tan M 2018 Investigation of the rheology and strength of geopolymer mixtures for extrusion-based 3D printing *Cement and Concrete Composites* **94** 307–14

[5] Zhang Yu, Zhang Y, Liu G, Yang Y, Wu M and Pang B 2018 Fresh properties of a novel 3D printing concrete ink *Construction and Building Materials* **174** 263–71

[6] Panda B and Tan M 2018 Experimental study on mix proportion and fresh properties of fly ash based geopolymer for 3D concrete printing *Ceramics International* **44(9)** 10258–65

[7] Sharanova A and Dmitrieva M 2019 Selection of compositions for additive technologies in construction *E3S Web of Conferences* **97** 06018

[8] 2018 *State standard GOST 8736-2014. Sand for construction work. Specifications* (Moscow: Standartinform)

[9] 2018 *State standard GOST 8735-88. Sand for construction work. Testing methods* (Moscow: Standartinform)

IOP Conf. Series: Materials Science and Engineering **684** (2019) 012002 doi:10.1088/1757-899X/684/1/012002

Simulation of the optimal control of a reagent supply in the in-situ uranium leaching process

M P Sadykov

L.N. Gumilyov Eurasian National University, System Analysis and Control Department, 2 Satpayev Street, Nur-Sultan, 010000, Kazakhstan

E-mail: sadykov.mahambetp@gmail.com

Abstract. This article develops an optimization task model for a physicochemical process that takes place in a mineral-bearing formation at the in-situ leaching of mineral. As a result of research, a certain hydrodynamic pressure and velocity field are obtained in a reservoir under the action network of wells, distributive reagent concentrations, mineral in a solid and liquid phase. Furthermore, according to this technique, the intensity-reagent impacts are studied on an extraction degree and mineral extraction time.

1. Introduction

The mineral extraction process using an in-situ leaching method (ISL) involves pumping a solvent into a mineral-bearing formation through injection wells, filtering a solution through an ore body with the selective mineral dissolution and its transfer from a solid phase to a liquid, and pumping the mineral-containing solution through pumping wells [1].

As a reagent in the leaching of uranium, sulfuric acid solutions, carbonate containing ammonium salts, sodium, potassium, etc. are used. The reagent choice depends on the mineralogical ore-deposits properties. In Kazakhstan, a sulfuric acid leaching has become widespread.

2. Mathematical and numerical models

The main reaction for describing the uranium leaching process with a sulfuric acid solution in a schematic form is [1], [2].

$$UO_3 + H_2SO_4 = UO_2SO_4 + H_2O.$$

This is explained by the good solubility of hexavalent uranium oxides in the sulfuric acid solutions, while the minerals of tetravalent uranium and coffinite under conditions of in-situ leaching practically do not interact with the sulfuric acid solutions without the oxidizing agents' addition to the solutions.

In an operational blocks-development by an ISL method, a reservoir is in the saturated state, therefore, a solution can be taken incompressible. The filtration process of incompressible solution (liquid) in a reservoir containing a dissolved mineral is described by the law of a mass conservation and filtration (Darcy's law) [3].

$$\operatorname{div} v = W, \tag{1}$$

$$v = -\frac{k}{\mu}\operatorname{grad} p. \tag{2}$$

Substituting an equation (2) into the equation (1) and replacing $h = p / \rho g$, it is possible to get an equation for a hydrodynamic pressure h in a reservoir in a form:

$$\frac{\partial}{\partial x}\left(K_{xx}\frac{\partial h}{\partial x}\right)+\frac{\partial}{\partial y}\left(K_{yy}\frac{\partial h}{\partial y}\right)+\frac{\partial}{\partial z}\left(K_{zz}\frac{\partial h}{\partial z}\right)=-W, \tag{3}$$

where ρ – fluid density;

K_{xx}, K_{yy}, K_{zz} – filtration coefficients (hydraulic conductivities), respectively, along with directions x, y, z;

W – sum of pumping and injection wells.

3. Discussion and results

Because of low uranium concentration in a reservoir, changes in the porosity and density of by-product due to the uranium dissolution with a reagent are neglected. The equations system describing mass transfer processes occurring during the uranium extraction by the in-situ leaching method is written as [4]:

$$\frac{\partial \bar{C}_m}{\partial t} = -\gamma \varepsilon C_R^0 \bar{C}_m \bar{C}_r, \tag{4}$$

$$\frac{\partial \varepsilon \bar{C}_r}{\partial t} = \operatorname{div}\left(\varepsilon D \operatorname{grad}\bar{C}_r - V\bar{C}_r\right) - v_1\gamma\varepsilon C_m^0 \bar{C}_m \bar{C}_r - \sum_d Q\delta\left(x_d, y_d, z\right)\bar{C}_r + \sum_p Q\delta\left(x - x_{0p}, y - y_0, z\right), \tag{5}$$

$$\frac{\partial \varepsilon \bar{C}_p}{\partial t} = \operatorname{div}\left(\varepsilon D \operatorname{grad}\bar{C}_p - V\bar{C}_p\right) + v_1\gamma\varepsilon C_m^0 \bar{C}_m \bar{C}_r - \sum_p Q\delta\left(x - x_{0p}, y - y_0, z\right)\bar{C}_p, \tag{6}$$

where

$$\bar{C}_m = \frac{C_m}{C_m^0}, \quad \bar{C}_r = \frac{C_r}{C_r^0}, \quad \bar{C}_p = \frac{C_p v_1}{C_r^0 v_2};$$

here V – filtration rate;

γ – coefficient characterizing a reaction rate;

C_m – uranium concentration in a solid phase;

C_r – concentration, sulfuric acid in a solution;

C_p – productive component concentration (uranium) in a solution;

Q – well flow rates ($Q_d < 0$ for pumping wells, $Q_p > 0$ for injection wells);

ε – reservoir porosity,

$v_1 = v_r R / v_m M$,

$v_2 = v_p P / v_m M$,

C_m^0 – initial mineral content in a reservoir;

C_m^0 – initial mineral concentration in a solid phase;

C_r^0 – reagent concentration on a pumping well;

$D_{i,j}$ – hydro-dispersion coefficient, determined by the following way:

$$\begin{cases} D_{xx} = \dfrac{\alpha_l u^2}{|V|} + \dfrac{\alpha_t v^2}{|V|} + \dfrac{\alpha_t w^2}{|V|} + D^*; \\ D_{yy} = \dfrac{\alpha_l v^2}{|V|} + \dfrac{\alpha_t u^2}{|V|} + \dfrac{\alpha_t w^2}{|V|} + D^*; \\ D_{zz} = \dfrac{\alpha_l w^2}{|V|} + \dfrac{\alpha_t u^2}{|V|} + \dfrac{\alpha_t v^2}{|V|} + D^*, \end{cases}$$

where α_l, α_t – longitudinal and transverse dispersion coefficients;

D^* – molecular diffusion coefficient.

Equations (4) - (6) are solved under the following initial and boundary conditions: at an initial moment, it is known that the mineral distribution in a reservoir, the concentration of a solution and dissolved useful component are absent:

$$C_m|_{t=0} = C_m^0,\ C_r|_{t=0} = C_r^0,\ C_p|_{t=0} = C_p^0,\ C_m|_S = 0,\ C_r|_S = 0,\ C_p|_S = 0.$$

When in-situ leaching, the reservoir-formation degree strongly depends on the input reagent concentration and the network location of wells [5]. Therefore, solving the optimizing tasks the mineral leaching process is reduced to solving two problems, one of which is the optimal control task of the flow rate of reagent and the other is the optimal network location of wells.

During leaching, when a well's location is already known, a process occurring in a reservoir can be controlled only by the input reagent concentration in an injection well. Therefore, in this paper, the reagent concentration C_r^0 is taken as a control. By controlling the reagent-concentration values C_r^0 in an injection well, a change in a mineral value in a liquid phase C_p^0 was investigated. In this case, a reagent concentration value set at an injection well should not exceed a certain maximum value. Accordingly, the optimal process control task is reduced to controlling a reagent supply rate in pressure well.

There are two restrictions on reagent concentration values input:

- Inequality limiting a reagent concentration at each injection well:

$$0 \le C_{rk}^0(t) \le C_{r\,max}^0; \tag{7}$$

- Total concentration equality of reagent at all injection wells to a certain value at each moment in time t:

$$\sum_k \int_0^T C_{rk}^0(t)\mathrm{d}t = C_R, \tag{8}$$

where k – number of injection wells.

An objective function for this task has a view:

$$f = \sum_d \int_0^T C_p(x_d, y_d, t)\mathrm{d}t \to \max, \tag{9}$$

where d – number of pumped wells;
C_{p} – mineral concentration in a liquid phase.

An algorithm for solving a problem is as follows: an initial approximation C_{r}^{0} is set, a direct problem is solved, which involves solving an equations system describing the mineral dissolution process from a reservoir when filtering a reagent-containing solution, in particular: an equation for the hydrodynamic pressure (3) is solved using an upper relaxation iteration method, a velocity field is determined from Darcy's law (1), the equations system for a mineral dissolution (4), the liquid-solution transfer (5) and dissolved mineral (6) is solved jointly by a Crank-Nicholson scheme. An objective function (9) is calculated from the found-values concentration of useful component.

Next, an inverse problem is solved, where the objective function gradient ∇f is calculated using a software package. Using the results of direct and inverse problems, an optimization problem (7) - (9) is solved, by means of which the optimal reagent-supply intensity values are determined at an injection well.

In modeling the uranium leaching with a sulfuric acid solution, the linear and hexagonal distribution of wells were considered. Due to the fact that linear wells are often used at the fields, a test calculation was carried out for a field block 20 meters wide and 40 meters long, consisting of three wells: two injection wells and one pumping (figure 1 (a)).

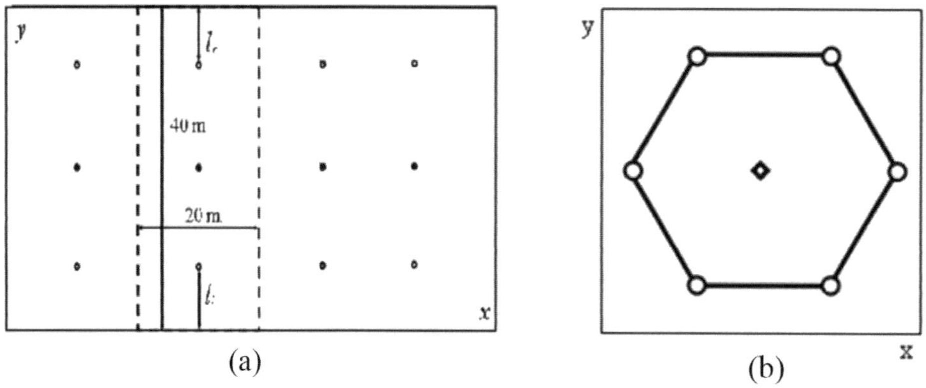

Figure 1. Schematic drawing of a considered area: (a) linear well network location in the case of 3 wells: at edges - pumping, in the middle - pumping out; (b) hexagonal wells network location: at edges - pumping, in the middle - pumping.

Figure 2 shows the pressure and isoline distributions of reagent concentrations, mineral in a solid phase and useful component in a liquid phase, at t = 100 days (vertical section along a well). Figure 3 demonstrates the degree-dependence formation of a reservoir and mineral concentration in pumped well overtime for the hexagonal location cases of wells. The scheme shows that at 70% recovery, the mineral concentration in a pumped well reaches a maximum value, and with the further mineral extraction through mining, the mineral concentration decreases. Figure 4 (a) and (b) illustrate the optimal reagent-concentration values in injection wells when constraints (7) and (8), respectively.

IOP Conf. Series: Materials Science and Engineering **684** (2019) 012002 doi:10.1088/1757-899X/684/1/012002

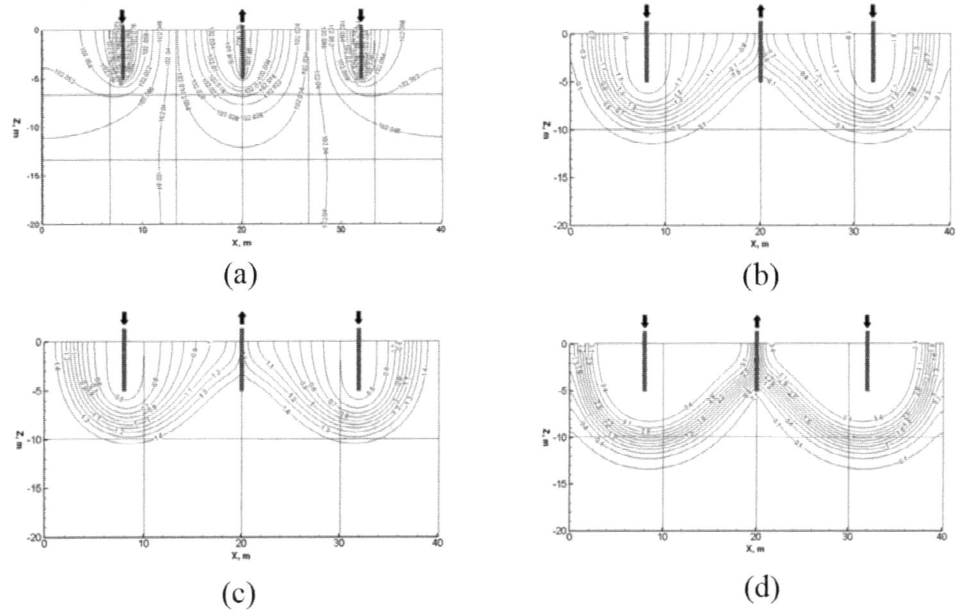

Figure 2. Pressure distribution and concentration isolines: a) pressure distribution, b) reagent, c) mineral in a solid phase, d) useful component in a liquid phase, with t = 100 days (vertical section along a well).

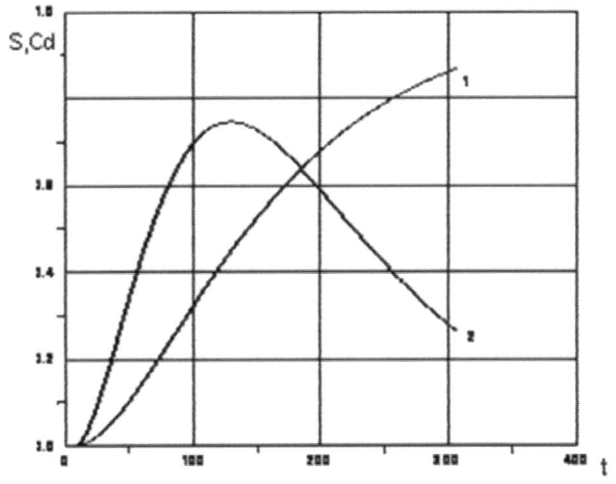

Figure 3. Dependence degree of reservoir formation and mineral concentration in a pumping well over time for the hexagonal location cases of wells: 1 - development field degree; 2 - relative mineral concentrations in a pumped well.

4. Conclusion

In this paper, an optimization problem model was developed for a physicochemical process that takes place in a mineral-containing formation at in-situ leaching. As a result of numerical studies, the obtained hydrodynamic pressure and velocity field were obtained in a reservoir under the action network of wells, the concentrations distribution of reagent, mineral in a solid and liquid phase. Obtained the optimal concentration values of the reagent in injection wells. The intensity influences of reagent supply on the extraction degree and mineral extraction time during the mineral extraction were investigated by an in-situ leaching method.

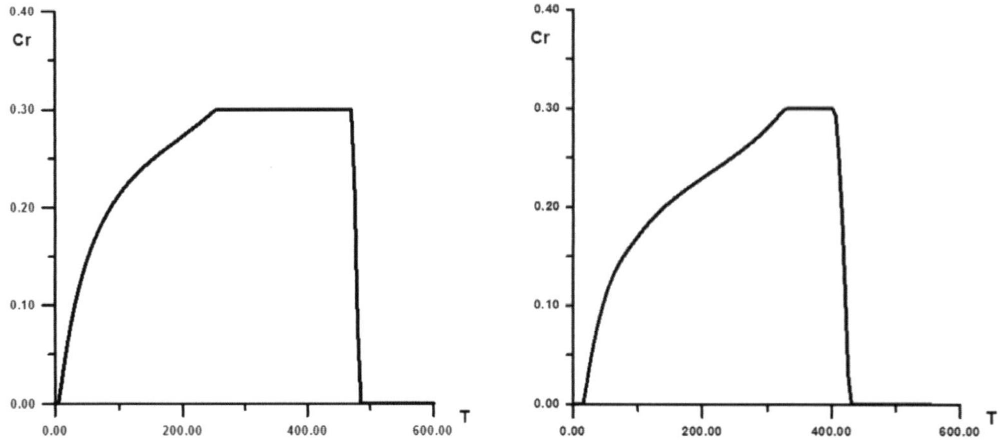

Figure 4. Optimal values concentration of the reagent in injection wells under different restrictions.

A created model for determining optimal flow regimes for a uranium mining can also be used for other mineral deposits developed by the in-situ leaching method.

References

[1] Bommer P M and Schechter R S 1979 *J. Society of Petroleum Engineers* **19** 393–401
[2] Parekh A C and Jung D H 1970 *J. Anal. Chem.* **42** 1423–27
[3] Santos I R, Burnett W C, Misra S, Suryaputra I G N A, Chanton J P, Dittmar T, Peterson R N and Swarzenski P W 2011 *J. Chemical Geology* **287** 114–23
[4] Hayworth H C, Ho W S, Burns W A and Li N N 2006 *J. Separation Science and Technology* **18** 493–521
[5] Nguyen V V, Pinder G F, Gray W G and Botha J F 1983 *J. Chemical Engineering Science* **38** 1855–62

The neural networks application in predicting the geometrical parameters of coatings formed on a steel substrate by laser alloying

N Ulakhanov and U Mishigdorzhiyn

East Siberia State University of Technology and Management, 40V Klyuchevskaya Street, Ulan-Ude, 670013, Russia

e-mail: nulahanov@mail.ru

Abstract. A mathematical model for predicting the geometric parameters of boron and aluminum-based coatings formed on a steel substrate by laser alloying is presented in the work. The model combines the principles of the theory of experiment planning with an artificial neural network (ANN) apparatus. Coatings were formed by means of laser heating of the steel surface with a pre-applied treatment paste. The ANN was built, trained and tested for the prediction of the width and depth of the molten bath on a steel substrate. The prediction results of the obtained neural network were compared with the corresponding linear models. The efficiency of the considered approach and the possibility of its application for predicting the values of laser treatment of steel parts of machines subjected to surface hardening are shown. The presented approach allows to use one mathematical model for predicting several parameters, thereby reducing the number of experiments.

1. Introduction

Nowadays combined processes of thermal-chemical treatment (TCT) and high energy fluxes have been effectively applied in surface engineering technologies of machine parts for surface (local) hardening [1-3]. In particular, laser radiation can modify a metal surface in a short time, compared to traditional TCT methods [1]. It is possible to obtain coatings with the same phase composition as traditional treatment, but different microstructure.

For practical application in order to build a computer system for predicting the results of laser reflow diffusion layers, interesting mathematical models that summarize experimental data, which are obtained in the form of a regression equation or in the form of power relations. Currently, one of the most effective methods of experimental data processing is the apparatus of artificial neural networks [4].

Artificial neural networks (ANN) is a mathematical model, as well as its software implementation, in which the process of transforming information is aimed at accumulating experimental knowledge and presenting it for further processing [4-6]. Interpreted information has a digital representation that allows to use ANN as a model of an object with unknown characteristics.

However, the capabilities of neural network models are limited by their learning method. The most well-known method of ANN teaching is the error propagation algorithm [4]. The disadvantages of this method are the unpredictability of the result as well as the duration of the learning process. The duration of training can be the result of a non-optimal choice of the step length and the size of the training sample, and the problem in learning usually occurs for two reasons: network paralysis and falling into a local minimum.

To improve the efficiency of the models based on the use of the mathematical apparatus of the ANN, it is important to reduce the number of experiments to obtain training pairs. The training set should adequately describe the process of diffusion hardening, i.e. contain maximum information about the effect of laser alloying parameters on the geometric parameters of the melting zone. From this point of view, it is of interest to use the results of mathematical experiments planning (MEP) or experimental design [7].

The purpose of this work is the ANN synthesis of predicting the geometric parameters of the coatings (width and depth of the molten bath) based on boron and aluminum, formed on a steel substrate using laser heating based on experimental data.

2. Regression models of the geometric parameters and the algorithm of the ANN application

The profiles investigation of the formed laser tracks on the surface of carbon steel showed that a two-layer structure is formed, consisting of a remelted zone (RZ) and a heat-affected zone (HAZ) (figure 1) [8].

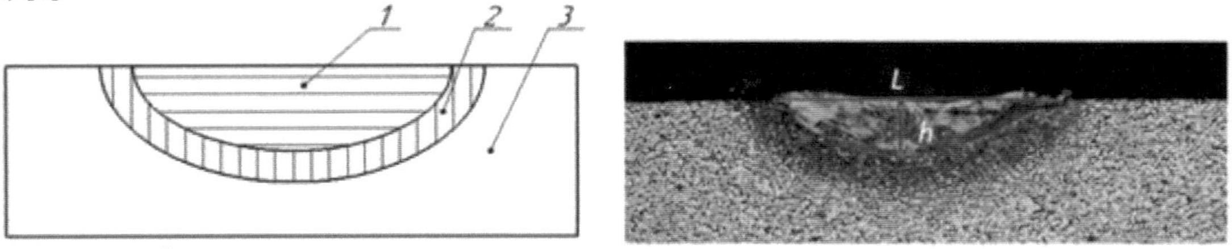

(a) (b)

Figure 1. The scheme of the laser alloying (a), where 1 is the RZ, 2 is the HAZ, 3 is the base metal (substrate); and macrostructure of the laser track cross-section (b), where *L* is the RZ width, *h* is the RZ depth.

HAZ is formed due to the rapid heat removal from the RZ to the substrate and is up to 50 μm thick. It will be not taken into account when the geometric parameters modeling of coatings.

The values of geometrical parameters (*L* and *h*) depending on the modes of laser heating are given in table 1.

Table 1. Geometric parameters of the RZ (μm).

Velocity *V* (mm min^{-1})	focus $Z = -2.9$ mm	focus $Z = -0.9$ mm
200	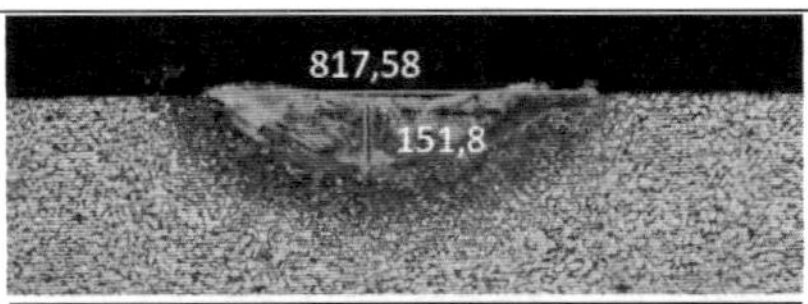	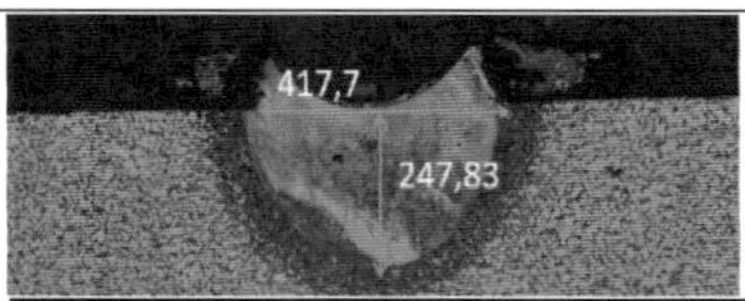
500	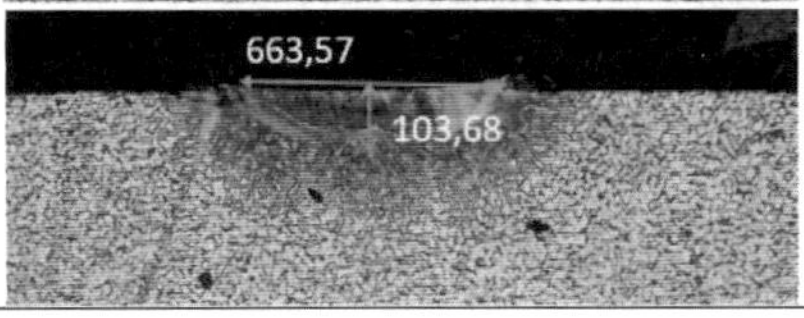	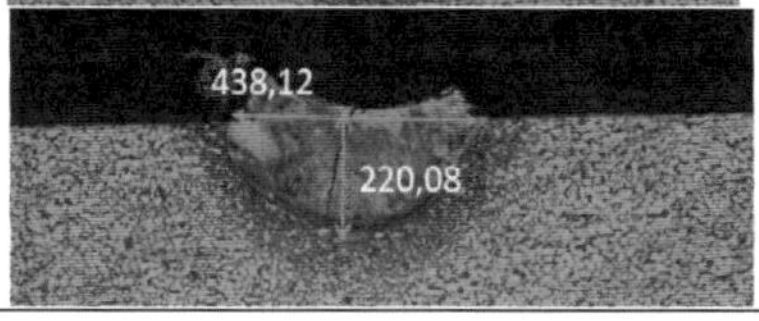

The initial data preparation and the construction of a mathematical model for predicting the geometric parameters dependence on the scan speed and focal length values were presented in previous research [8]:

$$h = 302.38 - 0.06 \cdot V + 40.7 \cdot Z + 0.04 \cdot V \cdot Z, \tag{1}$$

$$L = 242.46 + 0.19 \cdot V - 234.05 \cdot Z + 0.24 \cdot V \cdot Z. \tag{2}$$

All ANN data about the task is collected in a set of examples. Therefore, the quality of the neural network training depends on the number of examples in the training sample, to what extent these examples describe the task. The experimental results are being used as the training sample for both ANN and the construction of generalized mathematical models. To supplement the training sample, generalized models (1) and (2) are being used to increase the number of training pairs without increasing the number of experiments. Thus, the construction sequence of the ANN prediction consists of the following steps: 1) preparation the initial data; 2) the construction of a mathematical model using the MEP; 3) preparation of factor space for ANN learning on the basis of the initial data and the obtained polynomial mathematical model; 4) building a model of the ANN; 5) ANN training.

Two types of networks are suitable for solving the forecasting problem: a multilayer perceptron (MLP) and a network of radial basis functions (RBF) [4]. RBF networks provide more accurate results, but require a large amount of experimental data. With a relatively small dimension of the training sample, which is typical for this study, the MLP-type ANN is more effective. The main task at ANN building is to determine the number of neurons in the hidden layer. Figure 2 shows a graphical representation of the architecture of a synthesized neural network consisting of input, internal and output layers.

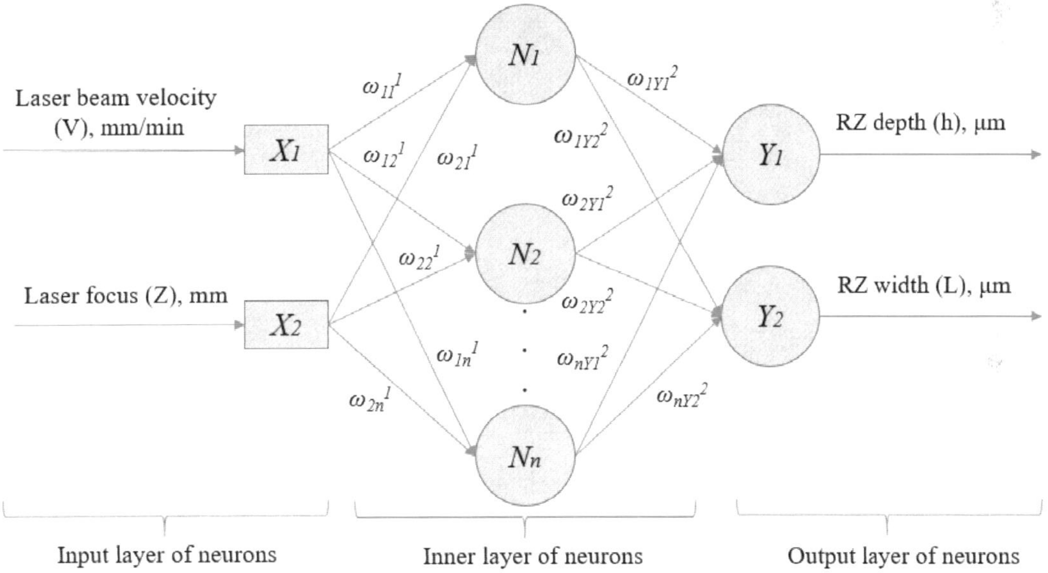

Figure 2. The topology of the ANN prediction, where: ω_i^k is the weight of the neuron connection of the previous layer with the next one.

ANN training was conducted using the error back-propagation method where its mathematical interpretation goes to the functional dependency [4, 5]:

$$F : F(X_i) \rightarrow y_i, i = 1, 2...132,$$

where X_i is the i-th row of the matrix of X input values; y_i is the i-th row of the matrix of Y output values (the i-th element of the column vector).

This is achieved by minimizing the objective function of the neural network error, calculated by the least squares method:

$$MSE(\{\omega_{ij}\}) = \frac{1}{2}\sum_{k=1}^{n}(r_k - y_k)^2,$$

where r_k is the value of the kth output of the ANN, y_k is the target value of the kth output of the ANN, n is the neurons number of the output layer, ω_{ij} is the connection weight linking neurons i and j.

3. Modeling of coatings geometric parameters using ANN

It is necessary to form training sets before building neural networks. The generated data matrix consisting of 50 examples obtained from equations (1) and (2) as well as previously conducted experiments will be used as a file of experimental data [8].

Table 2. Training and test sets for ANN.

No.	Velocity V (mm min^{-1})	Laser focus Z (mm)	RZ depth h (μm)	RZ width L (μm)
1	200	-2.9	149.15	820
2	500	-2.9	96.35	668.21
…	…	…	…	…
49	287	-1	232.98	462.16
50	230	-1	238.68	465.01

ANN modeling was performed using the Artificial Neural Network tool of Scilab package choosing the best network topology.

The simulation results are presented in figure 3.

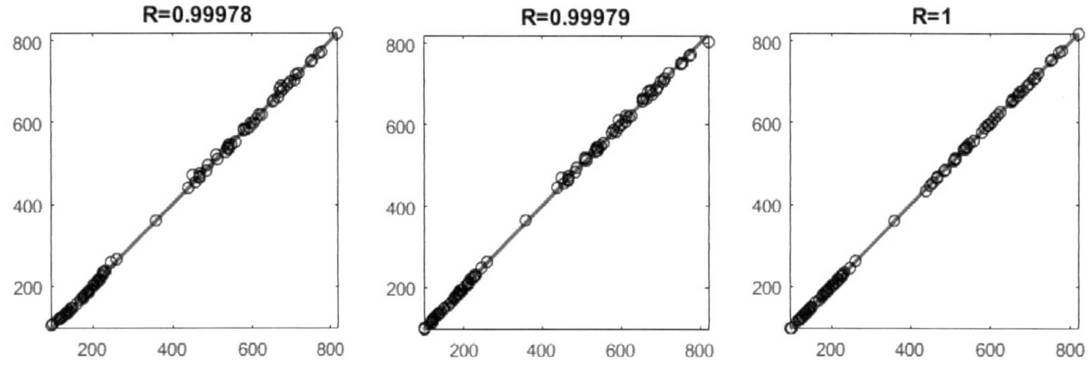

(a) (b) (c)

Figure 3. The regression coefficient R of the ANN training: (a) for a network with a 2-2-2 topology; (b) for a network with topology 2-5-2; (c) for a network with a 2-8-2 topology.

The best performance in terms of the correlation coefficient was shown by a network with the architecture of 2-8-2, where it equals to 1. There is an exact linear relationship between the ANN output and the target value at R = 1, if R is close to zero, then there is no linear relationship between these values [4, 5]. The results comparison of the calculations of the diffusion layer depth using the ANN and the model obtained using the MEP are presented in table 3.

Table 3. Comparison of the calculations results of the RZ parameters: width and depth.

Velocity *V* (mm min^{-1})	Focus *Z* (mm)	RZ depth *h* (μm) (MEP)	RZ width *L* (μm) (MEP)	RZ depth *h* (μm) (ANN)	RZ width *L* (μm) (ANN)
200	-2.9	149.15	820	149.16	819.24
500	-2.9	96.35	668.21	96.58	668.24
200	-0.9	247.41	447.2	246.57	448.3
500	-0.9	219.92	438.34	217.83	440.16

4. Conclusion

The ANN prognostic model based on experimental data and the results of MEP is able to distinguish functional relationships between the geometrical parameters of coatings and laser processing modes. The efficiency of this mathematical tool in the subject area is shown.

Acknowledgements

This work was supported by the grant of the President of the Russian Federation for the state support of young Russian scientists – PhD holders in 2018/2019 (Agreement no. 075-15-2019-514).

References

[1] Mishigdorzhiyn U L, Sizov I G and Polaynsky I P 2018 Formation of coatings based on boron and aluminum on the surface of carbon steels by electron beam alloying *Metal Working and Material Science* **20** 87–99

[2] Bartkowska A, Pertek A, Kulka M and Klimek L 2015 Laser surface modification of boronickelized medium carbon steel *Opt. Laser Technol.* **74** 145–57

[3] Podchernyaeva A 1997 Formation and properties of a surface layer during comprehensive laser boriding of carbon steels *Powder Metall Met. Ceram.* **36** 67–70

[4] Khaikin S 2008 *Neural Networks: a Full Course. 2nd ed.* (Moscow: Williams)

[5] Rutkovskaya D, Pilinsky M and Rutkovsky L 2006 *Neural Networks, Genetic Algorithms and Fuzzy Systems* (Moscow: Goryachaya Liniya—Telekom)

[6] Mebarek B and Keddam M 2018 A fuzzy neural network approach for modeling the growth kinetics of FeB and Fe_2B layers during the boronizing process *Matériaux & Techniques* **106** 603

[7] Adler Yu P, Markova E V and Granovsky Yu V 1976 *Planning an Experiment when Searching for Optimal Conditions* (Moscow: Nauka)

[8] Ulakhanov N S and Mishigdorzhiyn U L 2019 Study of the microstructure and modeling of geometric parameters of coatings based on boron and aluminum formed on a steel substrate by laser surfacing *Proc. XXX Int. Innovative Conf. of Young Scientists and Students (IICYSS - 2018) (Moscow)* (Moscow: Mechanical Engineering Research Institute of the Russian Academy of Sciences Press) pp 73–6

IOP Conf. Series: Materials Science and Engineering **684** (2019) 012004 doi:10.1088/1757-899X/684/1/012004

Processing of videos of the fatigue cracks propagation in the rotor blades of the helicopter

V Yu Kurokhtin

Strength of Materials Department, East Siberia State University of Technology and Management, 40v building 1 Klyuchevskaya Street, Ulan-Ude, Republic of Buryatia, 670013, Russian Federation

e-mail: kurokhtin91@gmail.com

Abstract. The article describes the processing of videos of growing fatigue cracks in the main rotor blades of a helicopter, which are formed during their fatigue tests on the stand. A specially created computer program allows to determine the geometrical parameters of a crack based on video recording (length, width, extreme points, center of gravity). Each frame of the video is binarized, after which a binary matrix is created on its basis, which is subject to element-by-element processing. The obtained data are presented in the form of a graph showing the dependence of the crack growth rate on the time of its growth. The graph is traced in the Mathcad program, as a result, two data vectors are created — the time and crack growth rate, which undergo an interpolation procedure with the construction of a third-degree polynomial. According to the results of the integration of the polynomial, the critical length of the crack is determined, after reaching which its rapid growth begins.

1. Introduction

During the fatigue tests of the main rotor blades of a helicopter, video footage of the growth of fatigue cracks arising in these blades is made [1]. An example of a sequence of frames from such a video is shown in figure 1, where the crack position is indicated by an arrow. The final frame corresponds to the moment of destruction of the sample - dividing it into two parts.

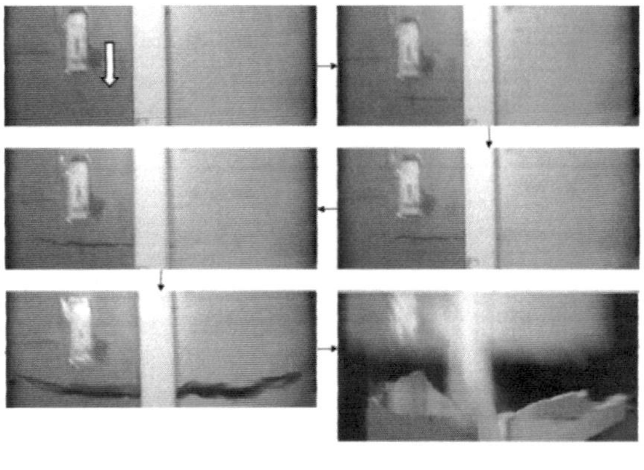

Figure 1. Images from crack growth video.

IOP Conf. Series: Materials Science and Engineering **684** (2019) 012004 doi:10.1088/1757-899X/684/1/012004

The aim of the study is to determine the growth rate of fatigue cracks, their critical length and time of subcritical propagation on the basis of available video.

To achieve this aim it is necessary to solve the following tasks: the creation of a method for processing video recordings; determination of the geometric parameters of the crack on each frame of the video based on the results of processing this frame; tracing and interpolation of the obtained data in order to determine the crack growth rate, time of its growth and critical length.

2. Research methodology

In order to determine the growth rate of cracks, a computer program was developed for video processing, in which each found crack has a separate entry in the internal database of the program. After the file was loaded into the program, the recording was divided into separate frames. Each frame was binarized (converted from color to black and white) [2]. The binarization threshold is selected so that the crack area on the frame remains black and all other pixels become white. An example of the image obtained as a result of binarization is shown in figure 2.

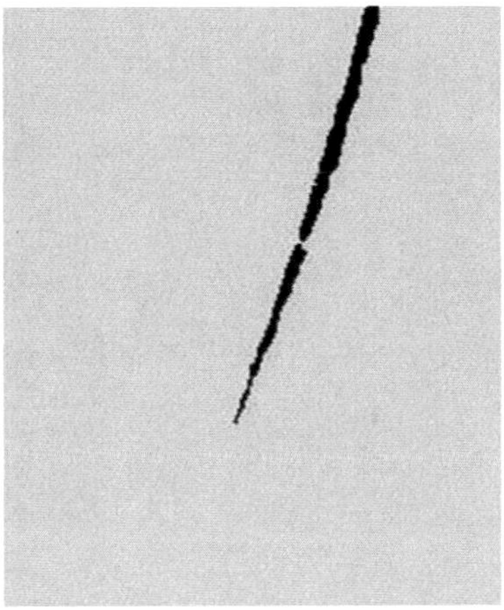

$$F = \begin{pmatrix} 0 & 1 & \overset{top}{0} & 0 \\ 0 & 1_{left} \leftarrow & \textcircled{1} & 0 \\ 0 & 0 & 0 & 1 \\ 0 & 1 & 1 & 0 \end{pmatrix}$$

Figure 2. Binarized frame and an example of a matrix corresponding to such a frame.

For each binarized frame, a matrix F was created consisting of zeros and ones (figure 2), where the white pixels of the background correspond to the zeros and the black pixels of the crack correspond to the ones [3]. This matrix was subjected to elementwise processing.

After processing the entire matrix, the database contains information about all the cracks found on the frame. These records were processed, the results of which identified the main parameters of the cracks - length, width, extreme points, center of gravity (figure 3).

Thus, following the results of processing the entire video, information was obtained on the dynamics of crack growth. These data are presented in the form of a graph (figure 4), where the abscissa axis is the time of crack growth in hours, and the ordinate axis is the rate of its growth in millimeters per hour [4]. It is seen that in the process of testing the crack growth rate gradually increases until a sharp jump occurs, ending with the destruction of the sample.

Small jumps of graphs up and down are due to the fact that the tip of the crack is poorly visible on some frames, so the rate of its growth is determined with some error - on some frames the crack looks longer, on others it is shorter.

The resulting graphs are traced in the Mathcad package (figure 5). As a result of this operation, two data vectors were created - the crack growth time in hours and the crack growth rate in millimeters per hour [5].

IOP Conf. Series: Materials Science and Engineering **684** (2019) 012004 doi:10.1088/1757-899X/684/1/012004

The data obtained were subjected to an interpolation procedure with the construction of the interpolation cubic polynomial. The interpolation cubic polynomial representing the dependence of the crack growth rate (mm h^{-1}) on the time of its growth (h) is described by equation (1):

$$v_{\text{length}} = 8.386 \cdot 10^{-7} t^3 - 1.364 \cdot 10^{-4} t^2 + 5.722 \cdot 10^{-3} t + 0.255. \quad (1)$$

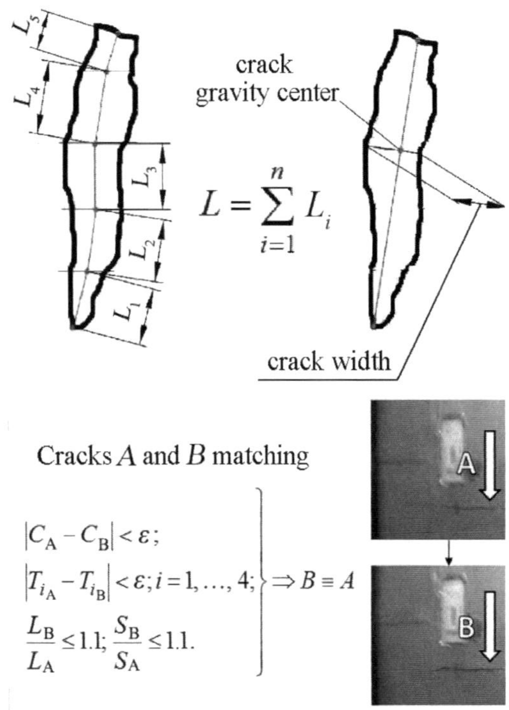

Figure 3. Determination of geometric parameters of cracks.

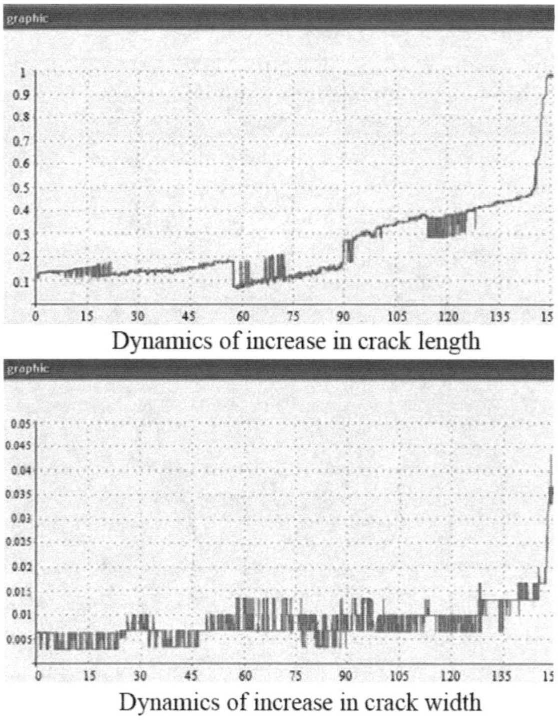

Figure 4. Kinetic graphs of crack propagation (horizontal axis – crack growth time (h); vertical axis – crack growth rate (mm h^{-1})).

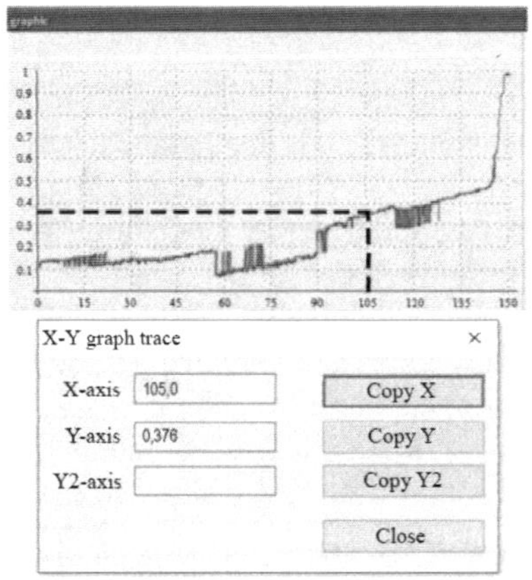

Figure 5. Tracing the graph in the Mathcad package.

IOP Conf. Series: Materials Science and Engineering **684** (2019) 012004 doi:10.1088/1757-899X/684/1/012004

3. Results

Substituting different values of crack growth time into formula (1), we obtain the values of crack growth rate. These data are presented in the table 1.

Table 1. Dependence of crack growth rate on the time of its growth.

Crack growth time (h)	Crack growth rate (mm h^{-1})
1	0.26
10	0.3
50	0.31
100	0.35
120	0.43
140	0.431
150	0.73
160	1.11
170	1.41
180	1.76
190	2.17
195	2.4
200	2.6

The data in the table is displayed as a graph (figure 6). As can be seen from the graph, the rapid growth of the crack begins after 140 hours of its propagation.

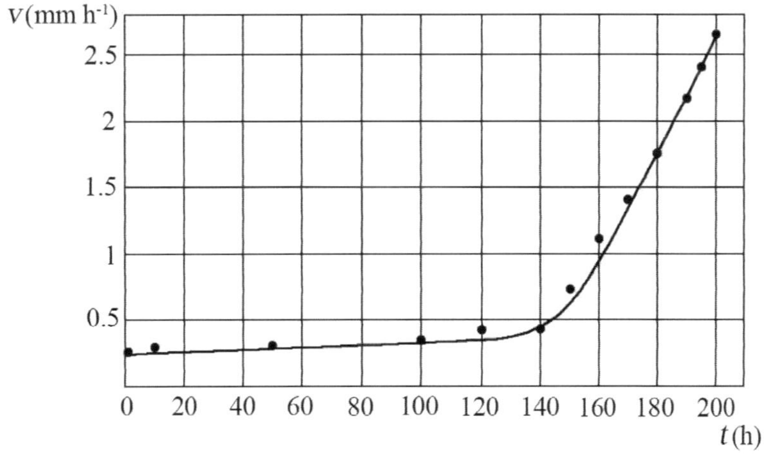

Figure 6. Graph of crack growth rate in blade samples depending on the time of its growth.

To determine the value of the critical crack length, we integrate equation (1) over time with an upper limit of 140 hours, corresponding to the beginning of the rapid crack growth:

$$\int_{0}^{140} 8.386\cdot10^{-7}t^{3} - 1.364\cdot10^{-4}t^{2} + 5.722\cdot10^{-3}t + 0.255\,\mathrm{d}t = 47.544\,\mathrm{mm}.$$

The obtained value is consistent with the experimental data - during the tests it was noted that rapid unstable crack growth begins when it reaches a length of 45-50 mm.

4. Conclusions

The period of subcritical crack propagation can be included in the assigned blade resource, since in this period, averaging the first 140 hours of crack propagation, the blade destruction does not occur. In a specified time, the crack grows to a length of 45-50 mm, after which its rapid unstable propagation begins, then the operation of the blade becomes dangerous.

The developed program allows to determine the growth rate and critical length of the crack with a sufficient degree of accuracy, which is confirmed by the good convergence of the results of the program with the experimental data. The developed software can be used to assess the growth rate and time of fatigue cracks in various engineering objects.

Acknowledgments

The work was supported by the state target of the Ministry of Science and Higher Education of the Russian Federation, project No. 9.11221.2018/11.12, and by the grant "Young Scientists of ESSUTM – 2019".

References

[1] Rogov V E, Bokhoeva L A, Kurokhtin V Yu and Baldanov A B 2017 *Naukovedenie* 80TVN217
[2] Kurokhtin V Yu, Bokhoeva L A and Kibireva E E 2017 *XVIII Rus. National Sci. and Tech. Conf. with Int. Participation "Mechanical Engineers to XXI Century" (Bratsk)* vol 16, ed P M Ogar, A A Suriev et al. (Bratsk: Bratsk State University Press) pp 240–4
[3] Kurokhtin V Yu, Rogov V E and Bokhoeva L A 2018 *Issues on Modern Machines Mechanics: Proc. of the VII Int. Sci. Conf.* vol 2, ed I G Sizov, V K Astashev et al. (Ulan-Ude: East Siberia State University of Technology and Management Press) pp 213–7
[4] Chermoshentseva A S, Pokrovskiy A M and Bokhoeva L A 2016 *IOP Conf. Ser.: Mater. Sci. Eng.* **116** 012005
[5] Potapov S D and Perepelitsa D D 2014 *Aerospace MAI Journal* **21-1** 104–10

IOP Conf. Series: Materials Science and Engineering **684** (2019) 012005 doi:10.1088/1757-899X/684/1/012005

Modeling and optimization of the processes of turbulence

I Ya Lvovich[1], Ya I Lvovich[2], A P Preobrazhenskiy[1] and O N Choporov[2]

[1] Information Systems and Technologies Department, Voronezh institute of high technologies, 73a Lenin Street, Voronezh 394043, Russia
[2] Information Security Department, Voronezh state technical university, 14 Moscow Avenue, Voronezh 394026, Russia

e-mail: komkovvivt@yandex.ru

Abstract. In this paper the problem associated with the modeling and simulation of turbulent flows is considered. Turbulent flows are quite common and a complete theory is not built for them at the moment. Among the models that are based on the subgrid scale, the model proposed by Smagorinsky was chosen. On the basis of the created subsystem in the form of a cross-platform dynamic library for modeling turbulent media, the simulation of water flow around a circular obstacle was carried out. The simulation results are presented. Suggestions are given for the optimization procedure, with variations in the implementation of the considered system of medium motion.

1. Introduction
Currently, it is possible to observe the development of integration of CAD (computer aided design, CAD) with modeling systems (Computer-aided engineering, CAE), allowing to test the created object directly during the design. In addition, the means to put computational experiments have been long used in science, when full-scale experiments are too expensive, impossible or necessary to be repeated a large number of times in the same conditions.

Despite the fact that most technical processes are described by known physical laws, it is often almost impossible to imagine in advance how an object will behave in certain conditions.

This fact is due only to the extreme complexity of modern systems. Computational physics is a way of conducting scientific and technical research, in addition to experiment and theory. In this case, the computer plays the role of a device that gives new opportunities to study the properties of various physical models.

2. The analysis of features of turbulent flows
The development of methods for numerical simulation of flows began in the 50s the last century with the advent of computational hydroaerodynamics. After the mid-70s to the 80s, due to the lack of computer power, computational fluid dynamics was used to a greater extent in scientific purposes, for the simulation of relatively simple and elementary cases. It was at this time that its mathematical and physical foundation was laid.

With the advent of the 90s to the present, the development of computer technology has allowed to widely apply the methods of numerical modeling.

Computational fluid dynamics has significantly expanded its mathematical and physical apparatus, improved models of turbulence, radiation, acoustics, two-phase flows, methods of discretization and numerical solution of differential equations, etc.

Thus, computational fluid dynamics is an applied science based on various fields of science and successfully used to solve a wide range of problems.

Currently it continues to develop actively.

With the development of computer technology and improvement of methods of calculation and data analysis, computational fluid dynamics can take a dominant place, surpassing the prevalence of experimental and analytical methods.

The advantages of numerical simulation of flows in engineering practice include:

- reducing development time of technical devices;
- modeling of processes and flows for which experiment is not available to conduct;
- detailed picture of processes;
- reducing the cost of development.

On the other hand, computational fluid dynamics has such disadvantages as the final accuracy of the calculation, which in some cases is capable of cross out the advantages of numerical modeling.

Along with the computational part, analytical and experimental methods are developed. The use of all three methods, mutually complementing each other, allows you to effectively develop complex technical devices such as aviation, space and automobile engines, aircraft, ships, submarines, ventilation systems, etc.

The simulation of turbulent flows is of a particular interest. Primarily due to the fact that such trends are most common.

In addition, there is still no complete theory that would describe similar flows. Research by experimental methods is also hindered. Therefore, the modeling of turbulent takes a leading role in practical calculations and theoretical studies.

3. The mathematical model

One of the largest branches of computational physics can be considered computational fluid dynamics (CFD).

Due to the application of approaches based on numerical modeling in the description of fluid motion, it is possible to significantly accelerate the development of different technical devices. Most movements of liquids, gas and plasma flows encountered in technology are turbulent [1].

A large number of models related to the subgrid scale are based on assumptions related to the existence of vortex viscosity [2].

There is a process of influence of small structures on large structures. This can be accounted for due to the introdused viscosity μ_t:

$$\mu_{\text{eff}} = \mu + \mu_t .$$

In cases where an unknown subgrid tensor τ_{ij}^{SGS} of stress τ_{ij}^{SGS} is considered, it is required to proceed from such a model equation [3]:

$$\tau_{ij}^{\text{SGS}} - \frac{1}{3}\tau_{ij}^{\text{SGS}}\delta_{ij} \approx \tau_{ij}^{\text{mod}}\delta_{ij} = 2\mu_t \overline{S_{ij}} .$$

The process of calculating the strain rate tensor using the expression:

$$\overline{S_{ij}} = \frac{1}{2}\left(\frac{\partial \overline{u_i}}{\partial x_j} - \frac{\partial \overline{u_j}}{\partial x_i} \right).$$

To date, a large number of models have been developed to determine the turbulent viscosity μ_t. It should be noted the model for subgrid stresses, which was suggested by Smagorinsky [4]. Similar to the Prandtl model, the turbulent viscosity is modeled as follows:

$$\mu_t = \ell^2 |\overline{S_{ij}}|.$$

Here $|\overline{S_{ij}}| = \sqrt{\overline{S_{ij}}\,\overline{S_{ij}}}$, and ℓ is the scale of turbulence. In order to determine the scale of turbulence, apply the width of the filter at the coefficient of proportionality C_S:

$$\ell = C_S \Delta.$$

In the above expression, the Smagorinsky constant is the coefficient of proportionality C_S. We can calibrate it. The range of its variation at different currents is between 0.01 and 0.24. We used $C_S = 0.1$ in the calculations.

The Smagorinsky model is characterized by simplicity and good convergence due to the fact that it is dissipative.

The considered model was supplemented by the procedure of multi-alternative optimization [5]. When multialternative aggregation was carried out by a variation of the embodiments of the system flow.

The vector of alternative variables was used on the basis of which the process of dichotomous reduction was realized.

The variety of sets of parameters was limited due to the fact that they were consistently divided into two parts and there was a choice of one of them as a starting point for the next division to be carried out.

In order to automatically implement the dichotomous reduction procedure, it is necessary to carry out the transition to the equivalent problem associated with a multi-alternative choice of a set of random variables.

They are independent random discrete numbers.

The reduction of the differences of the parameter sets will be performed in the case when the values of the probabilities of the possible realizations will be close to the value 1. For this purpose, it is required to record the optimization problem using alternative variables and to use randomized multi-alternative optimization when searching for solutions.

The general structure of optimization models reflecting the mechanisms [6, 7] under consideration is related to aspect of the duality principle implementation by dividing the efficiency indicators of the system into two groups that meet extreme and boundary requirements [8-10]. Full set of indicators $y_g, g = \overline{1,G}$. Some of these indicators $g_1 = \overline{1,G_1}$ are subject to extreme requirements

$$y_{g_1} \to \text{extr}, g_1 = \overline{1,G_1},$$

to the rest – boundary

$$y_{g_2} \le y_{g_2}^{\text{bd}}, g_2 = \overline{1,G_2}, \overline{1,G_1} \cup \overline{1,G_2} = \overline{1,G}.$$

These indicators depend on the choice of certain characteristics of the implementation of reduction and transformation mechanisms, which are formalized as a vector of optimized variables measured on continuous and discrete scales

$$x = (x_1, \ldots, x_j, \ldots, x_J),\ y_{g_1} = \Psi_{g_1}\left(x^1, x^2\right) \to \text{extr}, g_1 = \overline{1,G_1},$$

where $j = \overline{1,J}$ – the numbering set of optimized variables.

This is another aspect of the duality of the two forms of realization of the vector x:

- $x^1 = \left(x_1^1, \ldots, x_j^1 \ldots, x_{g_1}^1\right)$ – vector of continuous variables;
- $x^2 = \left(x_1^1, \ldots, x_j^2 \ldots, x_{j_2}^2\right)$– vector of discrete variables, in particular, Boolean:

$$x_j^2 = \begin{cases} 1, j = \overline{1, J_2}, \\ 0 \end{cases}.$$

The optimization model has the following form:

$$y_{g_1} = \Psi_{g_1}\left(x^1, x^2\right) \to \text{extr}, g_1 = \overline{1, G_1};$$

$$y_{g_2} = f_{g_2}\left(x^1, x^2\right) \le y_{g_2}^{\text{bd}}, g_2 = \overline{1, G_2};$$

$$x_j^{1\min} \le x_j^1 \le x_j^{1\max}, j = \overline{1, J_1};$$

$$x_j^2 = \begin{cases} 1, \\ 0 \end{cases}, j = \overline{1, J_2}.$$

Linear models with Boolean variables are of particular importance for system optimization

$$\sum_{j=1}^{J_2} c_j x_j^2 \to \text{extr};$$

$$\sum_{j=1}^{J_2} a_{g_{2j}} x_j^2 \le b_{g_2}, g_2 = \overline{1, G_2};$$

$$\sum_{j=1}^{J_2} a_{g_{2j}} x_j^2 \le b_{g_2}, g_2 = \overline{1, G_2};$$

$$x_j^2 = \begin{cases} 1, \\ 0 \end{cases}, j = \overline{1, J_2},$$

where c_j – coefficients of the linear objective function,

$a_{g_{2j}}$ – coefficients of the linear system of constraints.

The next aspect is related to two mechanisms of optimization modeling process organization: randomization and smoothing. Randomization is the transition from vectors x^1, x^2 to their probabilistic analogues. So instead of the coordinates of the vector $x_j^1, j = \overline{1, J_2}$ in the process of optimization of the search on the k iteration $k = 1, 2, 3, \ldots$ random realizations are considered $\tilde{x}_j^k, j = \overline{1, J}$, that meet the following condition for a given distribution type

$$m(\tilde{x}_j) = x_j^k,$$

where $m(\cdot)$ – the symbol of mathematical expectation. In the case of a vector of discrete variables and in particular Boolean variables, random variables are introduced $\tilde{x}_j^2, j = \overline{1, J}$ with distribution

IOP Conf. Series: Materials Science and Engineering **684** (2019) 012005 doi:10.1088/1757-899X/684/1/012005

$$P(\tilde{x}_j^2 = 1) = p_{x_j^2}, P(\tilde{x}_j^2 = 0) = q_{x_j^2}, p_{x_j^2} + q_{x_j^2} = 1 .$$

Introduction-Robin optimized random variables $\tilde{x}_j^2$, leads to another aspect of duality - dichotomy of reduction performed in automatic mode. The reduction of the variety of sets characterizing the training system is performed if the probability values of combining perspective subsets are close to 1, and unpromising subsets are close to 0. In the conditions of combinatorial uncertainty of the choice of perspective sets at the first step of dichotomous reduction the uniform distribution is accepted

$$p_{x_j}^1 = 0.5, q_{x_j}^1 = 0.5 .$$

4. Results of modeling

A subsystem was created in the form of a cross-platform dynamic library for modeling turbulent environments. It can be used in the development of CAD, as well as for scientific purposes. To check the correctness of the model, we created a simple geometry in the environment – water flow around a circular obstacle.

Then, based on the geometry, we will create a grid with the element size 0.5.

The figures below show the results of modeling turbulent water flow around an obstacle with Reynolds number equal to 100.

The length of the time step is 0.05, the simulation lasted for 2000 time steps, with a record of the result of the simulation every 10 steps.

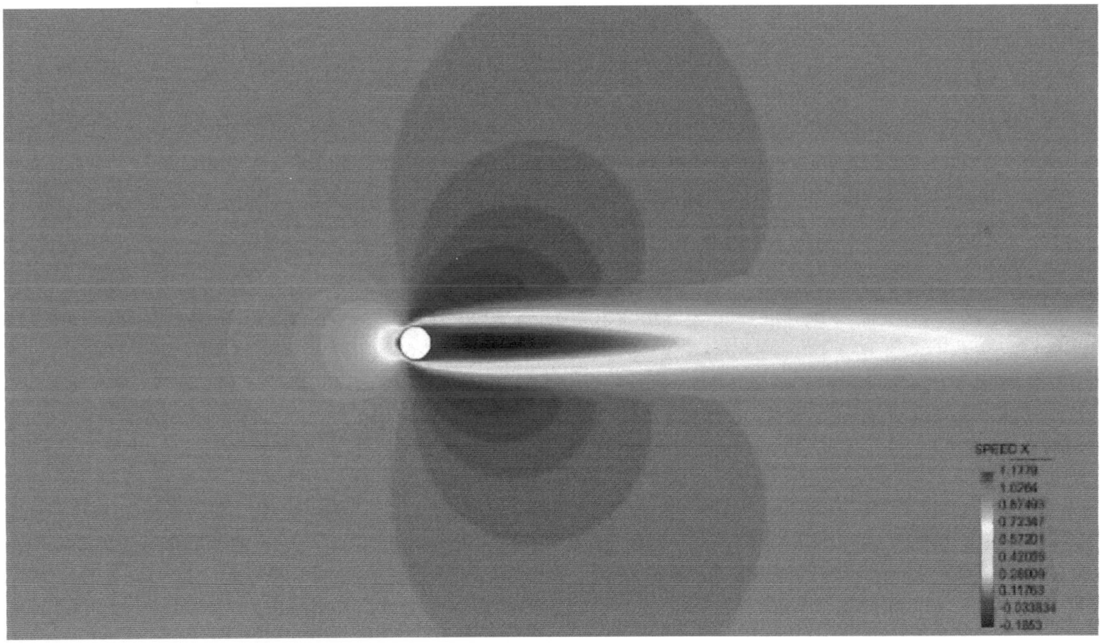

Figure 1. Step 80.

Tests of the program at various parameters of modeling have shown that the subsystem of modeling turbulent environments successfully copes with its task – simulates turbulent flow around various solids.

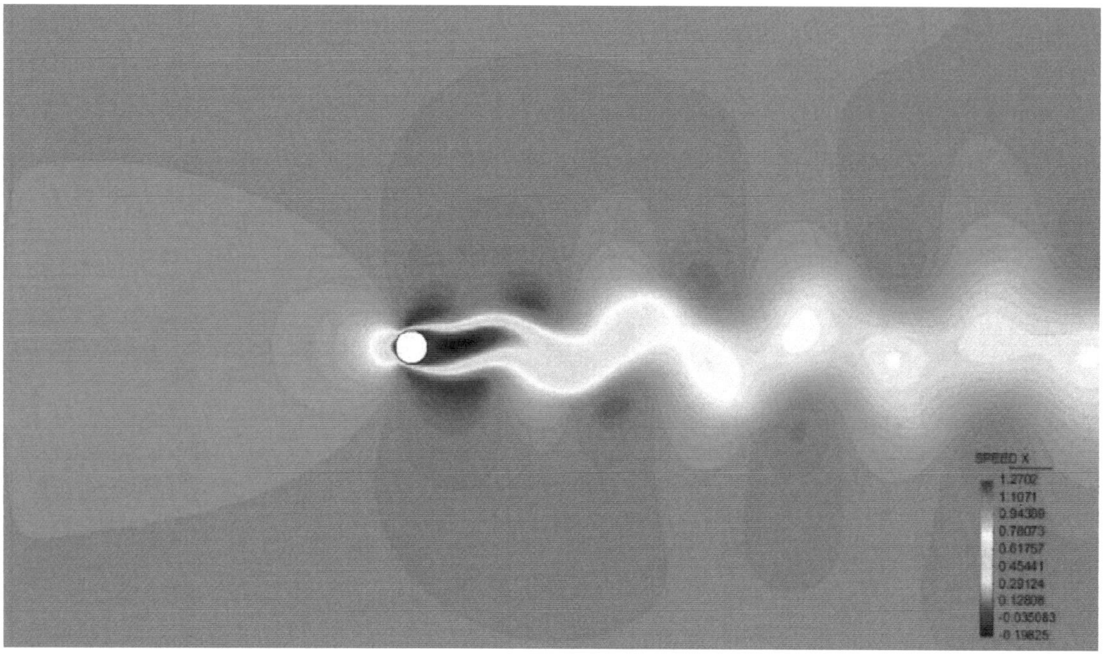

Figure 2. Step 200.

5. Conclusion

The algorithm for modeling turbulent environments on the basis of the Navier-Stokes equations and Smagorinski model was developed.

The algorithm was implemented as a compact cross-platform subsystem that can be used in the development of CAD, as well as for scientific purposes.

References

[1] Yun A 2005 *Development and Analysis of Advanced Explicit Algebraic Turbulence and Scalar Flux Models for Complex Engineering Configurations* (Darmstadt)
[2] Menter F R 2002 *Methoden, Moeglichkeiten und Grezen Numerischer Stroemungsberechnungen* (Erlangen: Numet)
[3] Wachter E M 2005 *Andwendung der Instationären Flamelet Methode auf Diffusions Flammen im Post-Processing-Modus* (VDI)
[4] Smagorinsky J S 1963 General circulation experiments with the primitive equations. 1. The basis element *Monthly Weather Review* **91** 99–164
[5] Lvovich Ya E 2006 *Multi-Alternative Optimization: Theory and Applications* (Voronezh: Kvarta)
[6] Ogawa N and Shimizu A 2016 Collegewide promotion of e-Learning/active learning and faculty development *Proc. of the Int. Conf. e-Learning* pp 179–84
[7] Igaki H et al. 2013 Programming process visualization for supporting students in programming exercise *J. Inf. Process. Soc. Jpn* **54** 1
[8] Kato T and Ishikawa T 2012 Design and evaluation of support functions of course management systems for assessing learning conditions in programming practicums *Proc. of ICALT-2012* pp 205–7
[9] Reschly A L and Christenson S L 2012 *Jingle, Jangle, and Conceptual Haziness: Evolution and Future Directions of the Engagement Construct. Handbook of Research on Student Engagement* pp 3–19
[10] Odu G O and Charles-Owaba O E 2013 Review of multi-criteria optimization methods - theory and applications *IOSR Journal of Engineering* **3** 1–14

Application of a discrete model of fatigue defect growth for assessment of the safe operation life of the main pipeline section with a corrosion defect

K Vansovich[1,3], **I Aistov**[1,4] **and A Nakhlestkin**[2,5]

[1] Omsk State Technical University, 11 Prospekt Mira, Omsk 644050, Russia
[2]JSC "Transneft – Western Siberia", 111 bldg. 1, Krasnyi Put Street, Omsk 644033, Russia

e-mail: [3]vansovichka@mail.ru, [4]aistov_i@mail.ru, [5]nakhlestkin95@mail.ru

Abstract. The article proposes a model of fatigue defect growth obtained on the basis of the analysis of the three-dimensional stress state at the apex of the corrosion defect. To study the stress-strain state the actual operational loads of the pipeline were simulated by the finite element method. The scientific novelty of the study is that when developing a model of fatigue defect growth, it is assumed that the growth of the defect, which develops by the opening fracture mode, is controlled only by normal stresses. In one loading cycle, the intensity of the fracture process is determined by both the maximum tensile stresses during the loading period and the maximum compressive stresses during the unloading period. Based on the assumptions made, it has been established that the growth rate of fatigue defects of corrosion origin depends on the magnitude of the coefficient of the average stresses change ahead of the crack front during full loading cycle.

To ensure non-stop operation of main pipelines, it is necessary to effectively assess the residual life of pipelines. During the operation of pipeline systems, structural changes in pipe steels occur [1]. This is due to the fact that the metal pipe is under load: static and cyclic. Under the influence of cyclic loads deformational aging of pipe steels occurs, which leads to reducing their resisting strength and brittle fracture.

It is accepted that:

A corrosion defect can be modeled as a crack-like defect, and the force pushing the crack can be calculated by means of elastic-plastic fracture mechanics [2].

In this paper, we propose a method for assessing the development of surface cracks and estimating the residual life of the main pipeline under cyclic loads, taking into account the biaxial stress state in the pipeline wall.

The object of study is the main pipeline with a crack-like defect of corrosion origin. The defect model was created on the basis of a real operational conclusion statement: a corrosion defect is, first of all, the loss of the pipe metal [3].

The annular stresses depend on the magnitude of working pressure in the pipeline; therefore, they are always tensile, i.e. positive [4]:

$$\sigma_{\text{annular}} = \frac{P \cdot D_{\text{in}}}{2\delta},$$

where P is pipeline pressure (Pa); D_{in} – pipe inner diameter (m); δ – pipe wall thickness (m).

Longitudinal stresses, depending on the temperature drop or bending direction, can be both tensile and compressive, i.e. negative. For a pipeline element clamped at both ends, the longitudinal stresses are determined as follows [5]:

$$\sigma_{long} = -\alpha \cdot \Delta t \cdot E + \mu \cdot \sigma_{annular} \pm \frac{E \cdot D_{out}}{2\rho_{eb}},$$

where α – temperature coefficient of linear expansion (1/°C); Δt – temperature differential (°C); μ – Poisson's ratio; D_{out} – pipe outer diameter (m); ρ_{eb} – radius of elastic bending of the pipeline axis (m); E – elastic modulus (Pa).

The development of fatigue defects occurs primarily due to the effect of annular stresses, which are the largest in magnitude and in their direction are perpendicular to the plane of growth of these cracks [6].

In this paper the bearing capacity of the pipeline is estimated by creating a mathematical model of corrosion defect using the ANSYS Workbench program. For this purpose, various loads that the pipeline undergoes during operation are simulated, taking into account the biaxial stress state.

To create the defect model shown in figure 1, the graphic editor ANSYS-Design Modeler was used (figure 2). The section of the pipe with a surface defect having a slight curvature.

Figure 1. Corrosion defect.

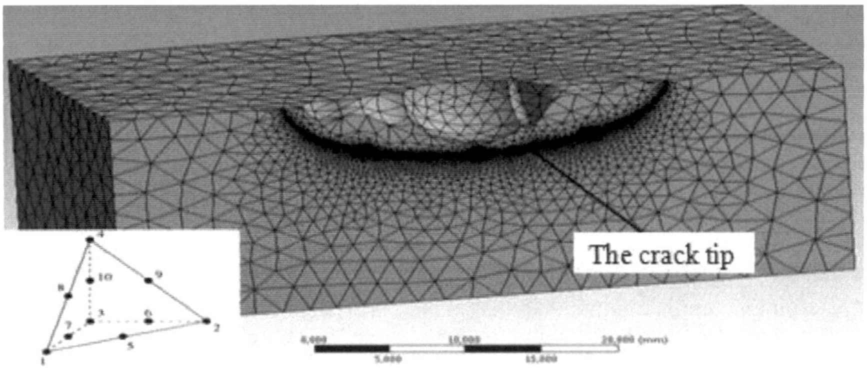

Figure 2. Corrosion defect model.

The experience of observing the growth of surface cracks indicates that they tend to take a semi-elliptical shape, which greatly simplifies the simulation.

After determining the principal stresses, it can be concluded that the largest values of the principal stresses are found at the points located in the area "lagging" behind the propagation of the crack front, which brings them in the process of growth to a semi-elliptical form (figure 3).

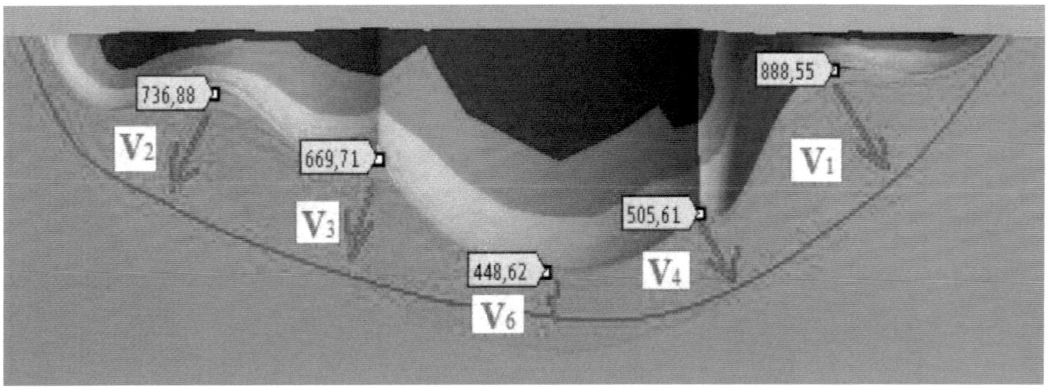

Figure 3. The semi-elliptical shape of the defect distribution.

The parameter Tr reflecting the change in the rate of growth of fatigue cracks under various types of biaxial loading of the pipe wall is the triaxial coefficient of the stress state [1]

$$Tr = \frac{\sigma_0}{|\sigma_0|},$$

where $\sigma_0 = \frac{\sigma_1 + \sigma_2 + \sigma_3}{3}$ is medium stress.

Given the triaxial coefficient, the equivalent stress intensity factor can be represented as:

$$K_{\mathrm{I}}^{\mathrm{Tr}} = Tr^{x} \cdot K_{\mathrm{I}}, \tag{1}$$

where Tr is the triaxial coefficient of the stress state; x is an exponent sign to be determined by the results of biaxial tests.

Taking into account (1), the expression for the rate of growth of cracks under various types of loading can be written similarly to the Paris formula:

$$\frac{\mathrm{d}a}{\mathrm{d}N} = C \cdot \left(\Delta K_{\mathrm{I}}^{\mathrm{Tr}}\right)^{n}. \tag{2}$$

In paper [1], a constant for steel 20 was defined as $C = 0.26 \cdot 10^{-10}$, and the equation for the growth rate of fatigue cracks, taking into account the nature of the stress state at its apex, will be as follows:

$$\left(\frac{\mathrm{d}a}{\mathrm{d}N}\right) = 0.26 \cdot 10^{-10} \cdot \left(Tr^{0.5} \cdot \Delta K_{\mathrm{I}}\right)^{4.0}.$$

The proposed model for the growth of fatigue cracks allows us to estimate the residual life of the pipeline in the presence of a surface or non-through crack. For this, equation (2) is integrated and the residual resource is determined $N_{r.1}$ i.e. the number of loading cycles from the moment of the crack registration l_{reg} to the critical size of the crack l_{cr}:

IOP Conf. Series: Materials Science and Engineering **684** (2019) 012006 doi:10.1088/1757-899X/684/1/012006

$$N_{\mathrm{r.l}} = \int_{l_{\mathrm{reg}}}^{l_{\mathrm{cr}}} \frac{1}{C_1 \cdot \left(\Delta K_{\sigma_0}\right)^{n_1}} \mathrm{d}a .$$

In this paper, a study has been conducted to assess the risk of defects of corrosive origin, as a result a mathematical model of pipeline fatigue defect has been developed; it allows to assess the development of identified defects, to estimate the safe operation life, and to impose restrictions on operating loads of the defective section of the main pipeline.

References

[1] Fokin M F, Erofeev V I, Smirnov S I and Sorokin G K 2005 *Applied Mechanics and Mechanical Engineering* (Nizhny Novgorod: Intel Service) pp 69–76

[2] Vansovich K A, Storks I P and Beselia D S 2018 *Engineering Journal: Science and Innovation* **5 (77)** 1–7

[3] Ostsemin A A and Zavarukhin V Y 1993 Strength of the pipeline with surface defects *Strength of Materials* **12** 1–59

[4] Pluvinage G, Boedra O and Melanin H 2018 *Science and Technology of Oil and Oil Products Pipeline Transportation* **4** 384–96

[5] 2011 *RD-23.040.00-KTN-115-11. Oil Pipelines and Oil Product Pipelines. Determination of Strength and Durability of Pipes and Welded Joints with Defects* (Moscow: JSC Transneft Press)

[6] Shiratori T, Miyoshi T and Matsushita Kh 1986 *Computational Fracture Mechanics* (Moscow: Mir)

IOP Conf. Series: Materials Science and Engineering **684** (2019) 012007 doi:10.1088/1757-899X/684/1/012007

Research on the technology of paste boronizing for H13 die steel

Zhang Yan-wei[1,2]**, Zhen Quan**[1,2]**, Burial Lygdenov**[1,2,3]**, A M Guriev**[1,2,4] **and Mei Shun-qi**[1,2]

[1] Hubei Digital Textile Equipment Key Laboratory, 1 Sunshine Avenue, Wuhan 430073, China

[2] School of Mechanical Engineering and Automation, Wuhan Textile University, 1 Sunshine Avenue, Wuhan 430073, China

[3] East Siberia State University of Technology and Management, 40v building 1 Klyuchevskaya Street, Ulan-Ude 670013, Russia

[4] Altai State Technical University, 46 Prospekt Lenina, Barnaul 656038, Russia

e-mail: meishunqi@vip.sina.com; 9859624870@qq.com

Abstract. In this paper, the boronizing effect of H13 hot work die steel is studied. The experiment is carried out to compare the boronizing effect of H13 steel at 850 °C, 900 °C and 950 °C. The heating time is set to 4 hours. The experimental results indicate that the temperature at 950 °C is suitable for boronizing of H13 steel. At 950 °C the boronized layer is more tightly bonded to the matrix, and the depth of the layer is deeper than at 850 °C and 900 °C. The hardness of the layer at 950 °C is better than at 850 °C and 900 °C in terms of the hardness of the layer. Therefore, 950 °C is a suitable temperature for boronizing the H13 hot work die steel.

1. Preface

Metal component surface treatments or surface modification techniques include such techniques as physical vapor deposition, chemical vapor deposition, induction beam hardening, hot dip coating, nickel plating, chrome plating, tin plating, carburizing, nitriding, carburizing, boronizing, and the like. Some of these treatments involve coatings that are the outer film layer of the metal substrate outer casing, while others process the physical and chemical properties over a distance of the substrate by diffusing another element. In the latter process, the change in substrate thickness is negligible. Boronizing is a surface diffusion treatment that increases some properties of the matrix to a certain depth [1,2,3]. After the boronizing treatment of steel material, the wear resistance, corrosion resistance and high temperature oxidation resistance can be improved, and the parts can be used in a more complicated and multi environment [4]. The performance of boronized steel is always superior to that of nitrided steel and carburized steel [5,6,7].

H13 hot work die steel is widely used in extrusion die and die casting. It is usually characterized by high strength and toughness. In direct contact with the work piece during operation, the temperature can reach 400-500 °C. In the forging machine, it can reach 650-700 °C. However, such die steels are commonly used in corrosive environments. The use of thermochemical treatment technology can improve the surface wear resistance, thermal fatigue and corrosion resistance. Boronizing technology is an effective surface treatment method. Due to the nature of the diffusion process, the boride layer has good adhesion to the substrate compared to conventional physical coating processes. Due to the high hardness of boride (1500~2000 HV), it has the advantage of high hardness compared with traditional surface treatment such as carburizing, nitriding and carbonitriding [8,9,10].

In this paper, the effect of temperature on the boronizing layer was determined by comparing the boronizing effect of H13 die steel at the same boronizing time at different temperatures. According to the literature [11,12], it can be concluded that the boronizing temperature is generally between 850 and 1000 degrees, so the temperatures are selected as 850 °C, 900 °C, 950 °C.

2. Experimental materials and experimental methods

2.1. Experimental materials

The experimental material used in this paper is H13 hot work die steel, the chemical composition of which is shown in table 1. The micro hardness of H13 steel is 450HV~550 HV, and the H13 steel is cut into 20 mm × 10 mm × 10 mm size, which is used for the micro hardness test of boronizing experiment. The experimental plasters are C, B_4C, NaF, and the like.

Table 1. Composition of H13 hot work die steel.

Element	C	Si	Mn	Cr	Mo	V	P	S
Content (%)	0.32~0.45	0.80~1.20	0.20~0.50	4.75~5.50	1.10~1.75	0.80~1.20	≤0.030	≤0.030

2.2. Experimental methods

The H13 steel sample, having a size of 20 mm × 10 mm × 10 mm, was sanded and polished with a sandpaper, and then the surface grease and the like were washed with alcohol and dried by a blower to be placed in a desiccator for use. After weighing the appropriate amount of the agent, add a proper amount of water to form a paste and apply it evenly on the sample. The thickness is about 3~5 mm. Place it in a dry and ventilated place and dry it during one night. Then put it in a dry box and heat it to 150 °C for 2 hours. Dry thoroughly. The heating temperature of the resistance box was set to three temperatures of 850 °C, 900 °C, and 950 °C, respectively, and three samples were heated, the heating time was set to 4 hours, and the sample was cooled to room temperature with the furnace.

After the sample is completely cooled, it is taken out and knocked out of the outer casing, and the sample is cut from the middle by wire cutting. The cut surface of the sample is polished, polished, etched, etc. with different types of sandpaper. The sample was placed under an optical microscope OLYMPUS DSX-HRUF to observe the layer structure. Hardness was measured by applying a force of 25 g on a (HV-100) hardness tester for 10 s.

3. Experimental results

3.1. Analysis of boronized layer

Three samples which had been subjected to boronizing treatment at 850 °C, 900 °C, and 950 °C were again polished and polished. According to the literature [13, 14], the surface of the treated sample was etched by disposing a 3 % to 5 % nitric acid solution, and then the boronized layer was observed under a microscope. The (a), (b), and (c) of figure 1 are boronized layers of 850 °C, 900 °C, and 950 °C, respectively.

It can be seen from (a), (b), and (c) in figure 1 that the boronized layer is sparse at 850 °C and the boronizing depth is shallow, when the temperature rises to 900 °C, the temperature of the boronizing layer begins to become tight and begins to appear jagged. When the temperature reaches 950 °C, the boronized layer is more compact, and the sawtooth shape is more obvious. The boronizing depth is obviously improved relative to 850 °C and 900 °C. Table 2 shows the depth of the boronized layer at

850 °C, 900 °C, and 950 °C. It can be seen that the effect of temperature on the boronized layer indicates that the boronizing effect at 950 °C is significantly better than at 850 °C and 900 °C.

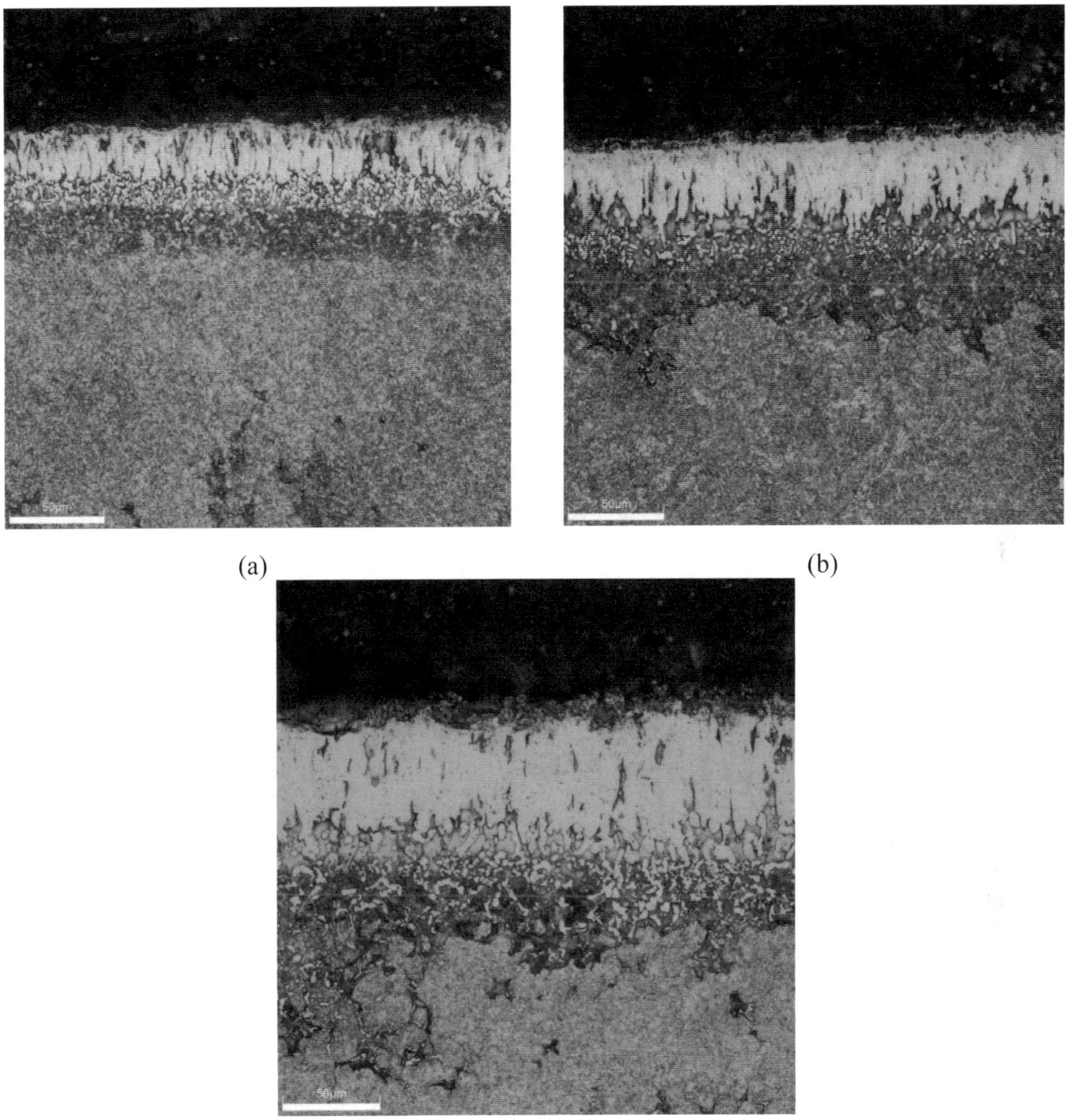

(a) (b)

(c)

Figure 1. The boronized layer at different temperatures.

Table 2. Depth of boronized layer of H13 steel.

Temperature (°C)	850	900	950
Layer thickness (μm)	30~40	35~40	55~60

3.2. Analysis of the hardness of the layer

Table 3 presents the hardness value of each point of the sample, and figure 2 shows the approximate hardness curve at 850 °C, 900 °C, and 950 °C.

Table 3. Hardness values.

	20 μm	40 μm	100 μm	200 μm	300 μm
850 °C	1127.82 HV	620.2 HV	475.10 HV	510.32 HV	463.33 HV
900 °C	1298.76 HV	570.3 HV	479.33 HV	530.20 HV	478.20 HV
950 °C	1402.77 HV	590.2 HV	454.32 HV	490.72 HV	530.21 HV

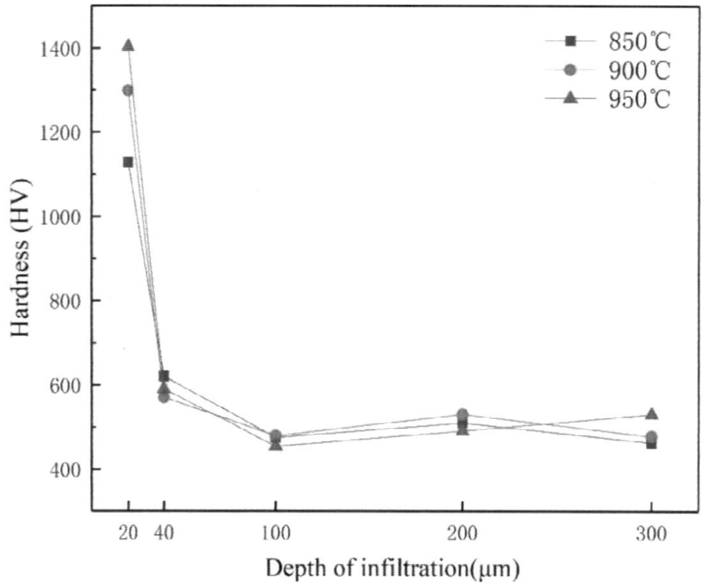

Figure 2. Hardness value curve.

Since only 20 μm of points on the thinner layer of the infiltrated layer are on the infiltrated layer, 40μm are on the transition layer, 100 μm, 200 μm, 300 μm are on the base material. It can be seen from table 2 and figure 2 that the hardness of the layer at 950 °C in the 20 μm layer is better than the hardness of the layer at 900 °C and 850 °C.

4. Conclusions

The research results indicate that the temperature at 950 °C is suitable for boronizing of H13 steel. At 950 °C, the boronized layer is more tightly bonded to the matrix, and the depth of the layer is deeper than at 850 °C and 900 °C. The hardness of the layer at 950 °C is better than that at 850 °C and 900 °C in terms of the hardness of the layer. Therefore, 950 °C is a suitable temperature for boronizing the H13 hot work die steel.

Acknowledgments

The work was supported by major science and technology projects in Hubei province (grant No. 2018AAA036).

References

[1] Özgen Ö S, Şeşen F E and Şeşen M K 2017 A study on boronizing kinetics of an interstitial-free steel *Materials Performance and Characterization* **6 (4)** 492–509
[2] Perry R H and Green D W 2008 *Chemical Engineering Handbook* (New York: McGraw-Hill)
[3] Davis J R 2002 *Surface Hardening of Steels: Understanding the Basics* (Ohio, Materials Park: ASM International Press)

[4] He Wei, Zhang Boping, Li Maolin, Tan Fuzhen and Lin Naiming 2016 Research status of boronizing process for iron and steel materials *Hot Working Technology* **45 (08)** 19–29
[5] Sundararajan G, Venkataraman B, Lakshminarayana K and Harish G B K 1998 *Studies on Boronizing of En 24 Steel* (Hyderabad: ARCI Press)
[6] Venkataraman B and Sundararajan G 1995 *Surf. Coating Technol.* **73** 177
[7] Jain V and Sundararajan G 2002 Influence of the pack thickness of the boronizing mixture on the boriding of steel *Surf. Coat. Technol.* **149(1)** 21–6
[8] Yang Zhe, Yang Haopeng, Wu Xiaochun and Qi Shengjun 2014 High temperature wear mechanism of double insulation solid boronizing of H13 steel *Chinese Journal of Materials Research* **28 (03)** 220–6
[9] Yang H et al. 2016 Enhanced boronizing kinetics and high temperature wear resistance of H13 steel with boriding treatment assisted by air blast shot peening *Surf. Coat. Technol.* **307** 506–16
[10] Campos I, Oseguera J, Figueroa U, Garcia J A, Bautista O and Kelemenis G 2003 *Mater. Sci. Eng.* A **352** 261–5
[11] Ma Zhuang, Li Ling, Wang Siqi, Wei Baoquan and Li Zhichao 2017 Research status of post-treatment of boronizing layer *Heat Treatment of Metals* **42 (08)** 146–51
[12] Xing Zhisong and Wang Baoping 2013 Test and application of boronizing process *Heat Treatment of Metals* **38 (08)** 99–102
[13] Liu Yumeng 2014 *Research on Boronizing Process of Self-Protecting Paste* (Nanjing: Nanjing Institute of Technology Press)
[14] Liu Jianjian and Chen Zhuping 2011 Pasting boron and its research status *Plating & Finishing* **33 (07)** 24–8

IOP Conf. Series: Materials Science and Engineering **684** (2019) 012008 doi:10.1088/1757-899X/684/1/012008

Study on the bending property of shirt sleeve slits

Gu Yi-Chuang[1,2] **and Mei Shun-Qi**[1,2]

[1] Hubei Digital Textile Equipment Key Laboratory, 1 Sunshine Avenue, Wuhan 430073, China
[2] Wuhan Textile University, 1 Sunshine Avenue, Wuhan 430073, China

e-mail: meishunqi@vip.sina.com

Abstract. Shirts are the clothing products that people need in all seasons. The folding effect of shirt sleeves directly affects the quality of the sleeves. In this paper by controlling the fabric of the shirt, the folding pressure and the additional pressure time, the fold recovery performance of the sleeves are studied from three aspects: bending stiffness, wrinkle recovery angle and bending recovery force. The experimental results show that the greater the areal density of the fabric is, the greater the restoring force is, the less likely the sleeves are folded. Under operation time of 5 minutes, the pressure of 30 N, 40 N, 50 N is applied to the wool fabric and the polyester fabric respectively, the restoring force was 1.0 cN, 0.75 cN, 0.5 cN and 1.2 cN, 0.8 cN, 0.6 cN respectively. At the same time the experimental results show that, the greater the additional pressure is, the smaller the resilience is, and the better the folding effect of the sleeves is. The research in this paper provides a theoretical basis for the actual production of the folding of the shirt sleeves and improves production efficiency by controlling the relationship between negative pressure and negative pressure time.

1. Introduction

With the improvement of people's life quality, people have higher and higher requirements on clothing. As a necessity of people's daily life, shirt is a kind of clothing favoured by consumers. People's demand for shirts is not only limited to wearing, but an aesthetic sense of the shirt is also increasingly important, today the market has a variety of styles, fashionable shirt products. A shirt belongs to the traditional classic clothing, attention to details and grasp in a men's shirt is particularly important. Sleeve slits are an important component of men's shirts, and their structure and technological form determine the quality and style of shirts [1]. With the development of science and technology, automation device has been used in all aspects of industry, but in the process of making sleeve slits, it faces many problems. For example, the production of armbands is low in automation and requires large labour costs. Even though some enterprises put in use the folding mechanism, it can only be applied to the folding of a single type of cloth and cannot be widely applied to the folding of cloth of different materials and thickness, which increases the production cost. In order to develop a folding mechanism suitable for different fabrics, it is urgent to study the bending properties of different fabrics. In this paper, the bending property of shirt fabric is examined on the basis of the study of shirt sleeve fabric.

The bending property of the fabric has a great influence on the appearance of the garment, its machinability and comfort. The bending property of fabric is an important mechanical property index for evaluating the style of fabric. The bending property of fabric is related to the bending property of fiber, the structure of yarn and fabric, post-finishing, etc. The relationship between them is very complex [2]. The methods used to study the bending properties of fabrics are analytical method, numerical method and rheological method. At present, the evaluation method of fabric folding and ironing effect is mainly subjective. The subjective evaluation method mainly evaluates the appearance of durable pressing products. Since it is a subjective evaluation, human factors, such as the observation proficiency of rating personnel, make the test results easy to produce bias, affecting the final results [3]. In this paper, the bending properties of different shirt fabrics are studied from the three aspects: bending stiffness, folding angle and bending restoring force; thus, the best folding effect of shirt sleeve fabric is found in order to choose the fabric better and achieve the best folding and ironing effect.

2. Bending rigidity of fabric

Researchers at home and abroad are committed to making objective evaluation standards for the bending properties of fabrics, putting forward many testing schemes and forming many bending theories, such as an inclined plane method, a heart method, a gray method, a cantilever method, a pure bending method and a saddle method. The inclined plane method mainly uses the mechanical model of cantilever beam to measure the bending length and calculate the bending stiffness, which has become the standard test method in various countries. The selection of experimental materials follows the principle of combining commonness, diversity and representativeness. According to the different thickness of fabrics and the different structure and organization, different groups of common cotton fabric, silk fabric and wool fabric are selected to be numbered respectively. Selected fabric basic parameters are shown in table 1 (fabric density units: gm^{-2}).

Table 1. Basic parameters of each fabric.

	Cotton	Cotton	Polyester	Polyester	Silk	Silk	Wool fabric	Wool fabric
Surface density	55.7	156.0	66.1	122.4	43.1	79.3	153.0	205.5
Tissue	Plain	Twill	Plain	Satin	Plain	Satin	Plain	Satin

According to GB/T 18318–2001 "Determination of the bending length of textile fabrics", the bending length of the sample in the standard atmospheric environment was tested by YG(B) 022D automatic fabric stiffening apparatus. The calculation method of bending stiffness can be expressed as follows [4]:

$$B = W \cdot C^3 \cdot 9.807 \cdot 10^{-6}.$$

In the formula, B is bending stiffness (cN cm^2 cm^{-1});
W is the fabric surface density (gm^{-2});
C is bending length (mm).

After multiple measurement and comparison the conclusion has been drawn: the fabric with the same material, the greater the surface density is, the greater the bending stiffness of the fabric is, the less easy it is to fold and iron; the same material fabrics, twill than plain when the bending stiffness; as a whole, the bending stiffness of mulberry silk is less than that of wool and less than that of the surface. According to the above results, the folding and ironing scheme can be appropriately adjusted when the sleeve vents of different materials are folded and ironed to achieve the best folding and ironing effect.

3. Fabric folding recovery angle

The crease recovery angle of the fabric is mainly used to judge the good or bad resilience of the fabric, and to test the anti-crease seam of the fabric [5]. However, in the opposite direction, it can also be used to judge the quality of fabric folding and ironing effect. Once the drape rebound angle of fabric is bigger, the flexibility of cloth is better, going against fold iron. There are two existing testing standards for fabric crease recovery in China, one is GB/T 3819–1997 "Determination of crease recovery angle method for textile fabric", which is an objective testing method, the other is GB/T 29257–2012 "Appearance evaluation method for textile fabric crease recovery", which is a subjective evaluation method. GB/T 3819–1997 is to keep the specimen folded along the warp (longitudinal) and weft (transverse) direction for a certain time, remove the load and recover for a certain time, measure the crease recovery angle, and use the measured angle to express the fabric fold recovery performance. The larger the rebound angle of the crease is, the better the crease resistance is. GB/T 29257–2012 is to apply a certain load to the torsion sample at a certain angle of three-dimensional torsion, so that it will produce wrinkles. The load will be removed and restored for a certain time, and then the degree of appearance wrinkling will be evaluated. Series is higher, fold is less, fight fold performance is better. Method 1 is more commonly used. In this paper, GB/T 3819–1997 "Determination of crease recovery angle of textile fabric" was used to identify the recovery angle of fabric folds [6]. In order to obtain a comprehensive result, several pieces of different fabrics with different thicknesses were selected, each material is shown in table 1 above, respectively to test the warp and weft bending performance and the warp and weft dry crease recovery angle (DCRA) of these fabrics. First sample cut into "product" glyph, in order to provide stability of the sample testing and representativeness, will be folded fabric according to the regulation, reply to wing sample size to 40 mm × 30 mm, and fixed on the wood, will reply wings folded, pressure on the regulation of the weight of a heavy hammer, after a certain time load is removed, unload load is measured with a protractor measuring sample reply wing back angle, 5 min after the crease recovery angle, the heavy hammer weight and loading time are recorded. The elastic recovery angle of longitude and latitude was measured 10 times respectively, and the average value of each direction was taken.

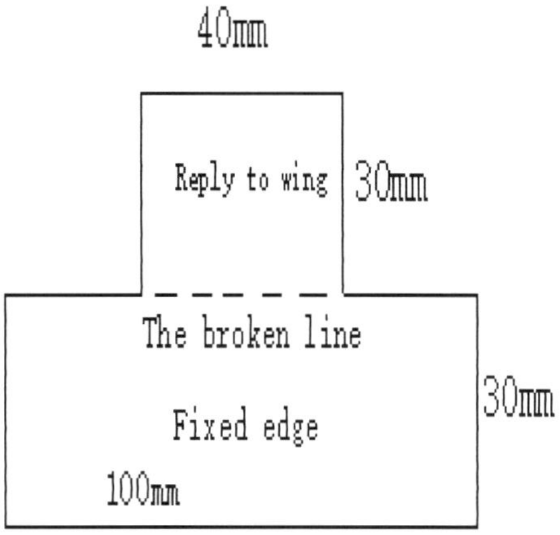

Figure 1. Sample cutting shape.

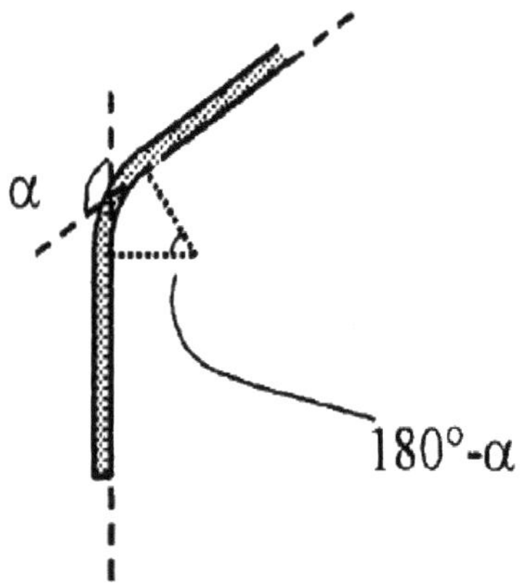

Figure 2. Fabric bending recovery angle model.

SAS is a famous data analysis software system in the world, and it is the first choice for data management and statistical analysis at home and abroad. We used the SAS system to conduct statistical regression by taking the warp and weft direction, bending property *B* and 2*HB* of the fabric as independent variables and its slow elastic wrinkle recovery angle as dependent variables. Through analysis and calculation, the regression equation of three independent variables is obtained [7]:

$$Y = 104.16293 - 10.35455 \cdot X_1 + 158.27376 \cdot X_2 - 235.39270 \cdot X_3,$$

where Y is the dry wrinkle recovery angle of slow elasticity (°);
X_1 is the longitude and latitude density (g km^{-1});

X_2 is the bending stiffness (cN cm^2 cm^{-1});
X_3 is the bending hysteresis moment (cN cm cm^{-1}).

According to the experimental observation results, by establishing a regression model with bending stiffness *B* and bending lag moment *2HB* as independent variables and slow elastic wrinkle recovery angle as dependent variables, it is concluded that there is a great correlation between them. The larger the bending stiffness is, the larger the crease recovery angle of the fabric will be.

4. The fabric bends to restore force

In the process of wearing and storage, the fabric is deformed due to external forces such as folding, compression and washing. Even after the external forces are removed, it is difficult to restore to the original flat state, which is called crease [8]. Although drape is an insurmountable problem in the appearance of clothes, it can be used in the production of sleeve slits. How to maintain drape, how much force to use and how long to achieve the best folding and ironing effect under negative pressure are also problems to be solved urgently. Many scholars have done a lot of research on the wrinkle problem, but most of the literatures focus on the theoretical analysis and discussion of anti-crease finishing and fold characterization of fabrics. When the fabric creases, the external force is much greater than the bending force, and the process of creasing and deformation also changes with time. Research make fabric drape and fabric bending force of bending force, both to test and to improve the fabric drape performance, better use of principle of fold, better wrinkle fabric production, and can give the fabric folded iron to provide theoretical basis for support, how to control the size of the force and effect time, achieve the best folding hot effect, make the fabric external appearance more beautiful. The parameters of the structure and bending performance of polyester fabric and wool fabric are shown in table 2. The experimental tools include weights of different weights, timers, protractor, force measuring devices etc. In order to control the independent variables and reduce the influence of other dependent variables, the experimental environment is the condition of constant temperature and humidity.

Table 2. Parameters of sample structure and bending performance.

Sample	Organization	The linear density (tex)	The density (article · (10 cm)$^{-1}$)		The surface density (gm^{-2})	The thickness (mm)	The bending stiffness (cN cm^2 cm^{-1})	
			radial	zonal			radial	zonal
Polyester	plain	26	224	193	161	0.0329	10.81	8.05
Wool fabric	plain	34	320	248	238	0.0618	16.58	13.53

Reference fabric crease performance test method standard of vertical method to test, in order to sample testing of stability and a representative sample, the sample reply wing is 40 mm × 30 mm size, and fixed on the block, the shape of the sample for the "product" word, will reply wings folded, pressure on the regulation of the weight of a heavy hammer, after a certain time after the removal of load, the chock with sample according to the rules set direction quickly in good homemade load device, sensor fabric from 3 mm, measuring the initial wrinkle recovery force from the fold. Under the condition of keeping the load action time constant, change the size of the load, measure for many times, and observe the size of the experimental restoring force.

In addition, the size of the load is maintained, and the duration of the variable action time is measured again to measure the drape recovery force of the sleeve slits of different materials, so as to better understand the influence factors and effect of the recovery force.

The result is shown in figure 3:

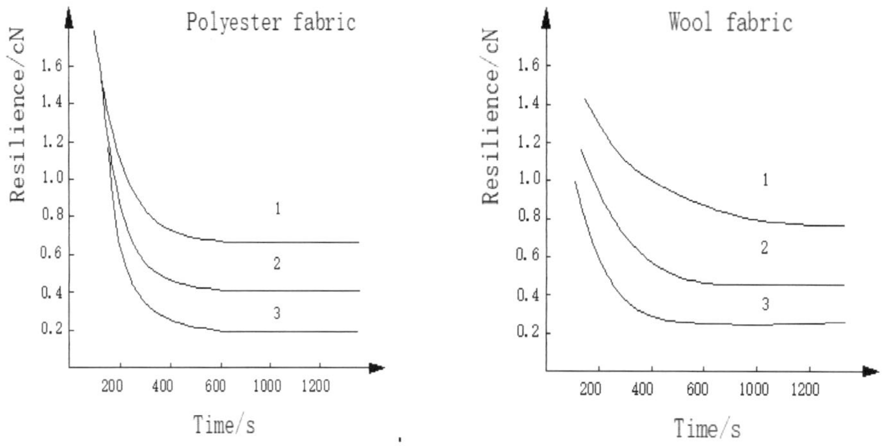

Figure 3. Relationship between recovery force and negative pressure and negative pressure time of different fabrics (curve 1 – 3 kg weight, curve 2 – 4 kg weight, curve 3 – 5 kg weight).

Observation and statistics of the final experimental results show that the pressure time will directly affect the wrinkle recovery force. For the same kind of fabric, the longer the pressing time is, the smaller the crease recovery force will be, and with the increase of time, the crease recovery force will start to decline rapidly and become gentle after a certain time. The different time of hammering will directly affect the size of the recovery angle of slow elastic wrinkle, for the same kind of fabric, the longer the pressing time is, the smaller the crease recovery angle will be. And with the increase of time, the crease recovery angle of the fabric with slow elasticity starts to decrease rapidly and becomes gentle after a certain time. Different fabrics under the same negative pressure and negative pressure time, the size of the recovery force is not the same.

5. Conclusions

Folding and ironing play an important role in the production of clothing. The folding and ironing of shirt sleeve slits is the process of making the cloth from one steady state to another steady state [9].

In this paper, cotton, wool, polyester, silk and other shirt fabrics of different density were selected for research. The research results show that:

- For cotton fabrics with different areal densities, the larger the areal density is, the larger the fold recovery angle is, which is not conducive to folding hot; choose different shirt fabrics, the greater the surface density is, the larger the fold recovery angle is, the folding effect is not obvious, which is not conducive to folding.
- The comparison of study results of two kinds of shirt fabrics: with polyester fiber and wool fiber, shows that under the same fabric and the same pressure action time, as the applied pressure increases, the resilience is getting smaller and smaller, and the folding effect is more obvious. Conversely, as the applied pressure is reduced, the resilience will increase and the folding effect will be poor. For the same material fabric, in case of a certain additional pressure, the longer the additional force action time is, the smaller the resilience is, the more obvious the folding effect is. In longer time period the resilience remains essentially unchanged and reaches a steady state.
- Under the same negative pressure and negative pressure time, the recovery force of the wool fabric is greater than that of the polyester fabric, because the surface density of the wool fabric is greater than that of the polyester fabric. Again, the greater the surface density is, the better the folding effect is, and the more favorable the folding of the fabric hot is.

This paper mainly studies the difference of three forms of expression under different negative pressures of different fabrics. It is concluded that the negative pressure is increased in production as much as possible, and the negative pressure time is shortened in order to improve production efficiency. There are also many influencing factors related to the folding and ironing effect, such as temperature, humidity, etc. In the follow-up study, the influence effect of various factors will be improved to show the best folding and ironing effect more accurately.

Acknowledgments
The work was supported by major science and technology projects in Hubei province (grant No. 2018AAA036).

References

[1] Yumei X 2008 Process optimization and application of men's shirt cuffs *Journal of Textile* **39(10)** 110–4
[2] Han L and Dexin H 2015 Discussion on shirt fabric *Liaoning Silk* **03** 18–9
[3] Yongshu M and Jibo W 2008 Design and research of automatic folding and pressing process for sleeve slits *Value Engineering* **37(32)** 169–70
[4] Qiaoying W, Ying H, Chunsheng W and Shihui L 2015 Comparison of test results of different fabric bending performance tester *Journal of Textile* **36(07**) 126–30
[5] Fengjun S and Jinlian H 2005 *Journal of Textile* **03** 15–8
[6] Haonan C 2019 Testing and index analysis of woven fabric's repeated bending performance *Wool Technology* **47(02)** 74–9
[7] Xiaorui T 2009 Study on the relationship between bending performance and wrinkle recovery angle of cotton fabric *Quality and Technical Supervision Research* **06** 25–37
[8] Yan W, Kan L, Runjun S and Meiyu C 2007 Test and research on wrinkle recovery of fabric *Wool Spinning Technology* **01** 54–7
[9] Jun L and Wenlin L 2014 Test analysis based on fabric crease recovery *Light Industry Standard and Quality* **01** 53–5

IOP Conf. Series: Materials Science and Engineering **684** (2019) 012009 doi:10.1088/1757-899X/684/1/012009

Complex saturation of corrosion resistant steels AISI430 and AISI303 with boron and carbon

Mei Shunqi[1], A M Guriev[1,2], S G Ivanov[3], M A Guriev[2] and B D Lygdenov[1]

[1] Hubei Digital Textile Equipment Key Laboratory, 1 Sunshine Avenue, Wuhan, 430073, China

[2] Altai State Technical University, 46 Lenin Avenue, Barnaul, 656038, Russia

[3] LLC «Hardening Technology», 39a Turinogorskaya Street, Barnaul, 656063, Russia

e-mail: serg225582@mail.ru

Abstract. Studies have been conducted on the effect of the degree of doping with chromium and nickel on the formation of boride coatings on steels AISI430 and AISI303. The saturation was carried out in a SNOL-type chamber furnace equipped with a PID controller from a powder saturating medium in an unsealed crucible. A mixture of boron carbide powders (70 wt. %), potassium tetraborate KBF4 (7 wt. %) and graphite – up to 100 % was used as a saturating medium. The process temperature was 900 °C, the saturation time was 2.5 hours. Upon completion of the high-temperature exposure, the crucibles were removed from the furnace and cooled in air to the 50–60 °C temperature. Diffusion coatings with a total thickness of 120 µm were formed on steel AISI430 and 75–85 µm on steel AISI303. The thickness of the boride layer for both steels was 45–50 µm.

Corrosion-resistant high-chromium steels are highly resistant to corrosion in aggressive environments. Therefore, they are used for the production of parts operating in corrosive environments. Since the high-alloyed steels contain a large number of alloying elements, the main of which are chromium and nickel, they also belong to expensive steels.

These factors are fundamental for maximizing the service life of parts manufactured from high-chromium corrosion-resistant steels. One of the probable ways to increase the service life of parts in the conditions of complex wear in aggressive environments containing a large number of abrasive particles is the method of boronizing. However, with an increase in the degree of doping, diffusion processes of the formation of coatings are significantly inhibited. One of the ways to intensify the formation of diffusion boride coatings is the simultaneous complex saturation of boron together with other elements [1–3].

In the present work, studies were conducted on the effect of the degree of doping with chromium and nickel on the formation of boride coatings on steels AISI430 and AISI303. The elemental composition of the heats of steel is presented in table 1. As this table shows, the elemental composition of these steels is almost the same, with the exception of nickel. The saturation was carried out in a SNOL-type chamber furnace equipped with a PID controller from a powder saturating medium in an unsealed crucible. A mixture of boron carbide powders (70 wt. %), potassium tetraborate KBF_4 (7 wt. %) and graphite — up to 100 % [3–5] was used as a saturating medium. All components of the saturating medium were ground in a ball mill to a fraction of 80–100 µm. The process temperature was 900 °C, the saturation time was

2.5 hours. Upon completion of the high-temperature exposure, the crucibles were removed from the furnace and cooled in air to the 50–60 °C temperature. Then the samples were removed from the saturating medium and washed in a warm soap-soda solution. The study of transverse microsections was carried out according to the procedures given in [6, 7].

Table 1. The elemental composition of the heats of steels involved in the work.

Steel	C	Si	Mn	S	P	Ni	Cr
AISI430	0.092	0.637	0.781	0.012	0.016	0.117	17.356
AISI303	0.098	0.722	0.877	0.013	0.011	9.863	17.981

The microstructure of the diffusion coatings obtained is shown in figure 1.

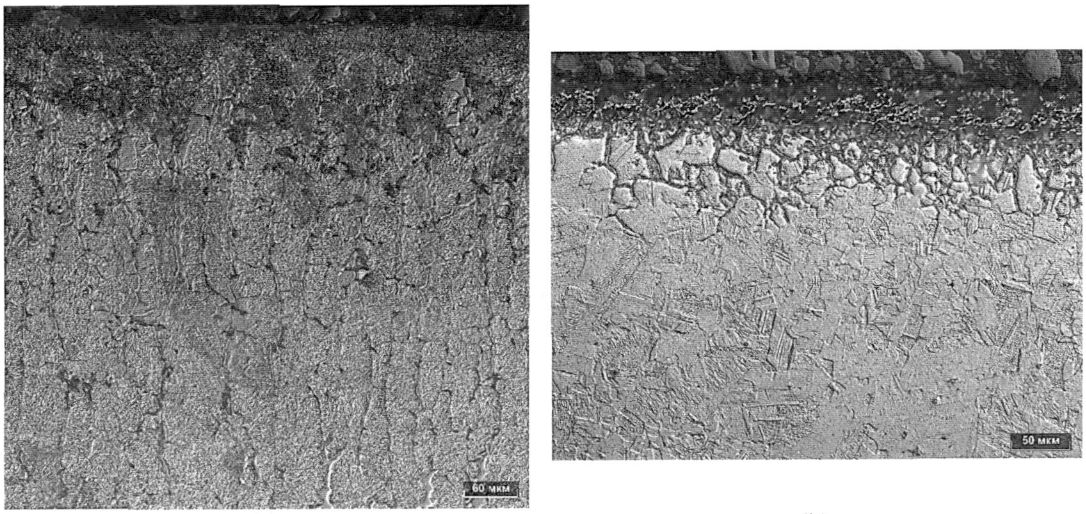

(a) (b)

Figure 1. The microstructure of the diffusion coating, obtained by simultaneous saturation of boron and carbon on steels: (a) AISI 430, (b) AISI 303.

Figure 1 presents the microstructures of diffusion coatings formed on both steels with total thickness: 120 μm on AISI 430 steel and 75–85 μm on AISI03 steel. The thickness of the boride layer for both steels was 45–50 μm.

Since AISI430 steel has a ferrite structure, and AISI303 steel has an austenite one and both phases are not etched, etching makes it possible to reveal the diffusion paths of boron and carbon along the corresponding phases, which are formed with their participation. According to [1, 3, 5], at boronizing high-chromium steels, chromium does not form its own borides, but dopes iron borides. Similar data are given for nickel. In the case of simultaneous saturation with boron and carbon, it is logical to assume that, along with borides, the formation of carbides and carboborides of complex composition is also possible. The main way of diffusion of boron and carbon active atoms is the grain boundaries, this is especially visible in figure 1(b). The layer of borides in the steels AISI430 and AISI 303 does not have a needle-like morphology characteristic of boride layers, but the interface is not smooth – there is a smooth transition from one zone to another, typical of diffusion coatings.

The distribution of microhardness over the cross section of diffusion coatings is shown in figure 2. As it can be seen from the graphs presented, the diffusion coating on AISI303 steel has a lower microhardness compared to AISI430 steel due to the effect of nickel.

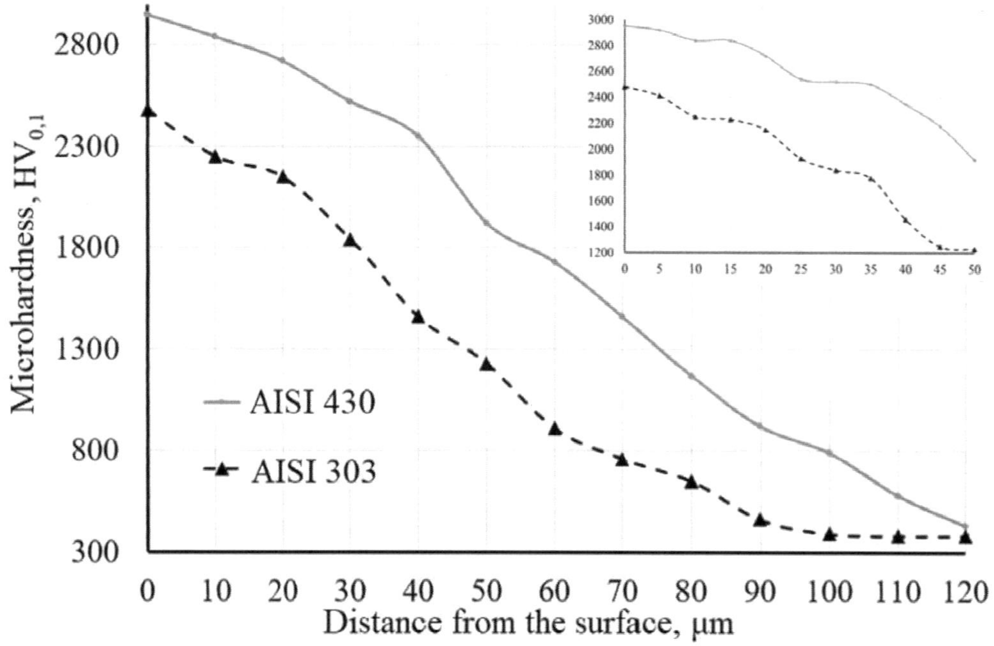

Figure 2. Distribution of microhardness over the cross section of diffusion coating on AISI430 and AISI303 steels after combined diffusion saturation with boron and carbon.

The graphs presented in figure 2 are the results of statistical data processing – each point of the main graph is an average of 15 measurements, each point on the incut is the result of 10 measurements.

References

[1] Kayali Y 2013 Investigation of the diffusion kinetics of borided stainless steels *Phys. Metals Metallogr.* **114** 1061

[2] Ivanov S, Garmaeva I, Guriev M, Guriev A and Starostenkov M 2015 Features of multicomponent saturation alloyed by steels *Lecture Notes in Mechanical Engineering* ed A Evgrafov (Cham: Springer) chapter Advances in Mechanical Engineering pp 49–53

[3] Dybkov V 2017 Boriding of chromium steel *Powder Metall. Met. Ceramics* **55** 650

[4] Guriev A, Ivanov S, Guriev M, Chernykh E and Ivanova T 2015 Chemicothermal treatment of tool materials *Steel Transl.* **45** 555

[5] Garmaeva I, Guriev A, Ivanova T, Guriev M and Ivanov S 2016 Comparative study of saturating power boriding media of different composition *Lett. Mater.* **6** 262–5

[6] Vander Voort G, Pakhomova O and Kazakov A 2016 Evaluation of normal versus non-normal grain size distributions *Materials Performance and Characterization* **5** 521

[7] Kazakov A and Kiselev D 2016 Industrial application of Thixomet image analyzer for quantitative description of steel and alloys *Microstructure. Microsc. Microanal.* **5** 294–301

IOP Conf. Series: Materials Science and Engineering **684** (2019) 012010 doi:10.1088/1757-899X/684/1/012010

Stretch elasticity and garment pressure of shaping-underwear fabric

Yun-Qi Gong and Shun-Qi Mei

Hubei Digital Textile Equipment Key Laboratory, 1 Sunshine Avenue, Wuhan, 430073, China

e-mail: meishunqi@vip.sina.com, 853573939@qq.com

Abstract. In recent years, the relationship between corsets and human health has been widely concerned. In order to explore the relationship between body shaping fabric and garment pressure, this paper starts from the reasons for the formation of body-dressing clothes, discusses the comfort range of human body wear and summarizes the principles, advantages and disadvantages of various methods of testing body-dressing clothes. At the same time, five elastic fabrics are selected to test their tensile properties. The garment pressures produced by the human calf diameter of 9 cm are also tested. The relationship between the tensile properties of the fabric and the garment pressure is analyzed, which lays the foundation for designing the intelligent corset that can meet the health requirements of people and realize the shaping of the body.

1. Introduction

Shaping-underwear, also known as tight garment, include bras, waist-nipper, split type shaping-underwear, toning pants, etc., mainly by applying pressure to the wearer to help people transfer excess meat from various parts of the body to the desired position. In this way, it can straighten the chest, lift the buttocks, slim the legs and correct the hunchback, so as to correct the body curve and shape the perfect figure for modern women [1, 2]. However, too much clothing pressure will not only cause the wearer's bone deformation, internal organs movement, but also bring them a series of physiological problems, and even cause diseases [3, 4]. Therefore, studying the pressure comfort of shaping-underwear is of great significance for optimizing its structural design and improving its fit and health.

2. Garment pressure produced by shaping-underwear

The garment pressure produced by the corset is mainly composed of weight pressure, surface pressure and restraint pressure. Weight pressure is formed by the weight of the body-shaping garment itself. Surface pressure is caused by the elasticity of the fabric and the looseness of the clothing or the excessive frictional resistance of the fabric when the human body is in dynamic contact with the garment. The restraint pressure is produced when the body is wrapped tightly in the body during exercise. The impact of the three on the human body depends on the specific situation. Generally, there are at most two of them in the plane of the human body, and sometimes only one kind of pressure exists. In the curved part of the human body, there may be two or three kinds of pressures. In addition, the cause of the corset garment pressure is also related to the psychological factors of the human body.

A large number of studies have shown that the pressure comfort on clothing required by the human body under normal exercise conditions is about 0.49 to 2.6 kPa. However, due to the different wearers and wear parts, the degree of discomfort will vary [5]. In 1995, Pratt proved through a large number of experiments that the blood pressure of the capillaries on the human body surface is close to 1.96~3.92 kPa. When the garment pressure reaches 5.88~9.8 kPa, people will feel uncomfortable. When the garment pressure exceeds 9.8 kPa, the human body is not able to withstand it [6]. When wearing a tight garment such as toning pants, if the leggings clothing pressure is less than 1.47 kPa, the human body basically does not feel the pressure. When the clothing is pressed at 1.47~2.46 kPa, the human body feels comfortable pressure. At this time, the exercise can be strengthened. When the clothes of the pants are pressed at 2.46~3.92 kPa, people begin to feel the pressure, but it is not enough to cause physiological harm. If the uncomfortable clothing pressure threshold reaches between 3.92 and 7.85 kPa, the wearer will feel extremely uncomfortable [7,8].

3. Test method for pressure of shaping-underwear

3.1. Subjective testing method

The evaluation of garment pressure produced by corsets has special characteristics. Shapewear will wrap the body tightly, so the comfort level is directly perceived by the body itself. Therefore, the subjective evaluation method is often used to study the clothing pressure. In order to quantify subjective feelings, psychological scales are often used in trials to assign different levels of sensation to different values, allowing the wearer to mark the scale based on their wearing sensation. At the end of the test, the tester estimates the size of the value based on the marker. However, due to the large difference in subjective evaluation of individual feelings, simple subjective test cannot accurately describe the comfort performance of clothing. So, it is necessary to use instruments to obtain more objective results [9].

3.2. Objective testing method

Objective tests include direct measurement and indirect measurement. The direct test method is to place the sensor in some parts of the corset and the wearer in the dressing state, and directly measure the pressure value of the specific part. The direct test method is divided into many types, as shown in table 1 [10, 11].

Table 1. Several direct garment pressure testing methods.

Test Methods	Principle	Advantages	Disadvantages
Fluid pressure test	Measuring the garment pressure value by reading the height change of the mercury column in the single tube or the height difference of the liquid level in the U-tube.	Simple operation, light weight and small size.	Low data reliability, without continuous dynamic measurement.
Sensor measurement	Including piezo-resistive type, capacitive strain type, piezoelectric type and pressure change detected by voltage or resistance value change.	Light weight, small size, high precision, stable results, real-time measurement of continuous pressure distribution.	Easy to be affected by factors such as radius of curvature, hardness, and elastic elongation of the fabric.
Air bag pressure tester	By applying pressure to the airbag filled with a certain amount of air, the pressure sensing member transmits the pressure to the input end of the strain gauge pressure	Good applicability, small error and good flexibility.	Only point measurement can be performed, and local pressure distribution law cannot be predicted.

IOP Conf. Series: Materials Science and Engineering **684** (2019) 012010 doi:10.1088/1757-899X/684/1/012010

	sensor, and the change in pressure is detected as a change in voltage.		
Elastic fiber pressure test	The incident light is generated by the He-Ne laser generator, enters the optical fiber and is reflected back to the core from beginning to end. The elastic fiber is deformed by force to change the number of rays transmitted in the core layer. And the external force value of the elastic fiber can be obtained by the amount of light.	Simple structure, small size, being commonly used for socks, bandages and other ribbon fabric testing.	Being restricted by pressure measurement site.
FlexiForce Stress Test	Through the FlexiForce pressure sensor, the measured physical signal is converted into an electrical signal by the amplifying circuit, and then converted into a computer by the data acquisition system, and finally processed, analyzed and displayed by the computer.	Excellent technical characteristics and easy to use with continuous measurement and analysis.	It is suitable for static measurement and the cost is high. Dynamic test can also be carried out when the measurement accuracy is not high.
Virtual instrument based pressure test system	The signal collected by the sensor is processed and sent to the PC through the signal conditioning device and the data acquisition card. Finally, the data is analyzed by the Labview program.	It improves the measuring efficiency, realizes multi-point simultaneous measurement and real-time observation of pressure changes.	When measuring, the sample content is large, and it is necessary to master high theoretical knowledge of computer, physics, mathematics etc.

Differences in the laboratory environment, temperature, humidity, and the physiological condition of the wearer will have a certain impact on the direct test results. In order to effectively avoid these interference factors, indirect measurement methods can be used. It is also divided into many types, as shown in table 2 [12, 13, 14].

Table 2. Several indirect garment pressure testing methods.

Test Methods	Principle	Advantages	Disadvantages
Theoretical calculation method	Based on the establishment of the three-dimensional human body model, the garment pressure is theoretically calculated from the elongation deformation and curvature in all directions.	Various effects can be simulated to predict the pressure distribution.	Pressure calculation results have large errors.

Plaster method	Gypsum or synthetic resin is used to form a convex model that simulates the elbow, knee, etc. Holes are punched in the arch, and a pressure sensor is attached to measure the pressure on the raised part of the clothes.	The natural garment pressure can be measured at the time of dressing.	The gypsum model material is hard and far from the softness of the human body.
Soft dummy method	The shape of the soft dummy is digitally scanned by a humanoid scan. And it has a special skeleton, soft tissue, smooth skin and a pair of easily changing artificial breasts.	The modulus of elasticity is similar to that of a real person, with a wide range of applications, saving time and cost.	Ignoring the effects of physical and psychological reactions on the body's stress.

4. The relationship between fabric and garment pressure

As a kind of clothing with adjustment function, the corset is generally made of elastic fabric. Compared with ordinary fabric, elastic fabric has the mechanical characteristics of easy elongation, easy recovery and good elasticity. The elasticity of the fabric depends on the stretching of the fabric, while the garment pressure is caused by the compression of the fabric into the body. Therefore, the garment pressure must be related to the elongation of the fabric.

4.1. Basic parameters of the fabric

In order to make the test results representative, five representative corset fabrics were purchased from the market, and their parameters are shown in table 3.

Table 3. Fabric parameters.

Fabric number	Sample 1	Sample 2	Sample 3	Sample 4	Sample 5
Composition and content	5% spandex + 95% polyester	8% spandex + 92% polyester	12% spandex + 88% nylon	15% spandex + 85% nylon	18% spandex + 82% nylon
Gram (g m^{-3})	180	176	198	192	170

4.2. Fabric extension performance test

When the human skin is in motion, the maximum elongation of the longitudinal and transverse skin is 45 % and 34 %, respectively. Therefore, during normal wearing, the elongation of the garment material is similar to the skin elongation and will not exceed 50 %. In this paper, the longitudinal and transverse tensile properties of elastic fabrics were tested using an electronic universal testing machine. The tensile curves of the five fabrics are shown in figure 1 and figure 2.

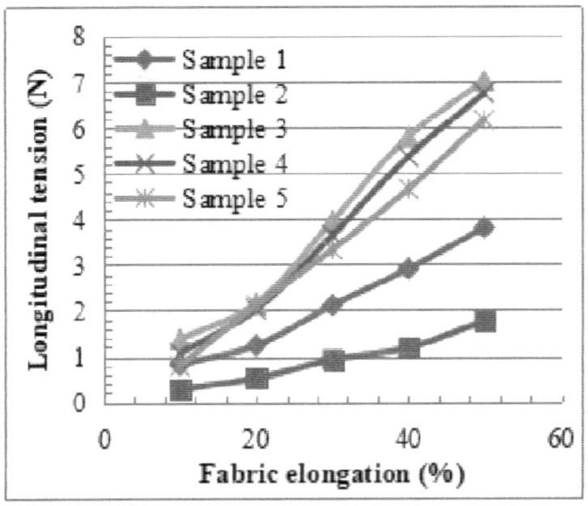

Figure 1. Longitudinal tensile curve of the fabric.

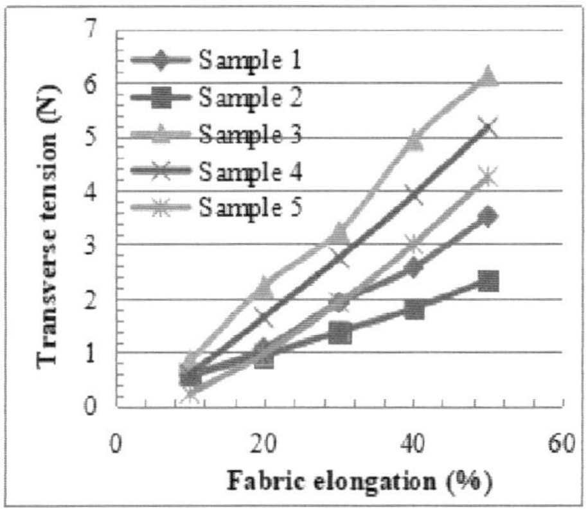

Figure 2. Transverse tensile curve of the fabric.

During the stretching process of the five fabrics, the tensile forces in both longitudinal and transverse directions increased with the increase of the elongation. Moreover, in the initial stage of stretching, the sample can produce a large elongation deformation under the action of a small tensile force.

4.3. Research on garment pressure of body shaping fabric
In this paper, the pressure of the garment with a diameter of 9 cm on the human calf was measured with a sensor. Through curve fitting, the relationship between the garment pressure generated and the fabric elongation within the range of 50 % was obtained, as shown in figure 3 and figure 4.

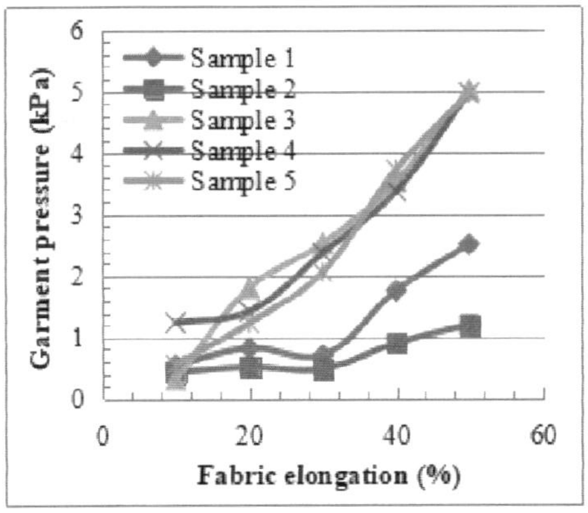

Figure 3. Longitudinal elongation and pressure curve of fabric.

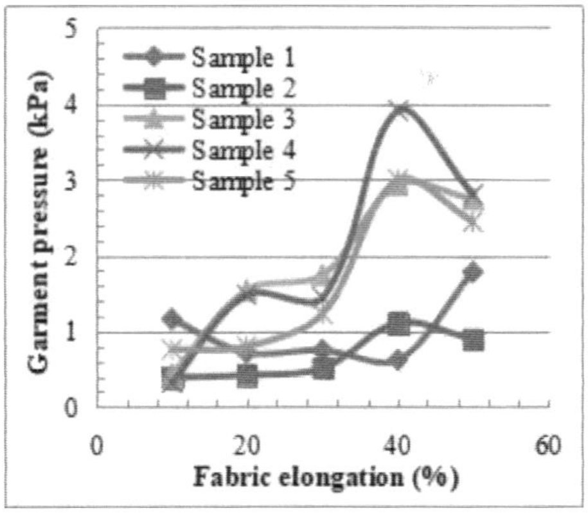

Figure 4. Transverse elongation and pressure curve of fabric.

Except for sample 1, during the longitudinal stretching of the fabric, the garment pressure increases linearly with the tensile elongation, which may be caused by either the different spandex content of the fabric or the different fabric structure. Compared with the transverse stretching, the longitudinal stretching has a greater impact on the garment pressure generated by the human body, which may be because the garment pressure is mainly caused by the longitudinal stretching of the fabric.

5. Concluding remarks

The paper starts with the reasons for the garment pressure on the body moulding underwear and summarizes the principles, advantages and disadvantages of various methods of testing body shaping clothes. At the same time, this paper also focuses on the relationship between the extension of some elastic fabrics and the garment pressure. This kind of discussion is limited to the surface. It is also necessary to experiment with more elastic fabrics to study the specific relationship between fabric and garment pressure.

Acknowledgments

The work was supported by major science and technology projects in Hubei Province (grant no. 2018AAA036).

References

[1] Zhou J, Wang A X and Liu S S 2018 Research on shaping effect of shaping-underwear *Journal of Xi'an Polytechnic University* **32(02)** 152–7

[2] Liu X H 2017 *Research on Comfort System Evaluation of Body-Shaping Underwear Based on Neural Network* (Beijing: Beijing Institute of Fashion Technology Press)

[3] Guo Z R, Xu J and Liu X T 2012 The test experiment of corset on human physiological effects *Journal of Xi'an Polytechnic University* **26(02)** 168–73

[4] Gaied I, Drapier S and Lun B 2006 Experimental assessment and analytical 2D predictions of the stocking pressures induced on a model leg by medical compressive stockings *Journal of Biomechanics* **39(16)** 3017–25

[5] Wang Q, Chen D S and Wei Q F 2009 Situation and prospects of the researches on clothing pressure effects on body *Journal of Textile Research* **30(04)** 139–44

[6] Pratt J and West G 1995 *Pressure Garments: a Manual on Their Design and Fabrication* (UK: Bath Press)

[7] Makabe H et al. 1993 Effect of covered area at the waist on clothing pressure *Sen i Gakkaishi* **49(10)** 513–21

[8] Liao L X 2012 *Research on the Loose Design of Body Shaping Underwear* (Suzhou: Suzhou University Press)

[9] Chen D S, Liu H and Wang L Z 2012 Clothing pressure test technology based on safe dressing and advances in its evaluation research *Advanced Textile Technology* **20(04)** 53–8

[10] Lee Y and Hong K 2013 Development of indirect method for clothing pressure measurement using three-dimensional imaging *Textile Research Journal* **83(15)** 1594–605

[11] Hu Y Q, Zhang H and Deng Y M 2015 Progress in the research of health field-oriented clothing pressure *China Textile Leader* **5** 70–2

[12] Ishimaru S et al. 2011 Prediction method for clothing pressure distribution by the numerical approach: attention to deformation by the extension of knitted fabric *Textile Research Journal* **81(18)** 1851–70

[13] Ma L et al. 2018 Research on finite element analysis contacted mechanical model of bra and human breast based on 3D digital measurement *Shandong Textile Science & Technology* **59(04)** 27–9

[14] Yu W, Fan J and Qian X 2004 A soft mannequin for the evaluation of pressure garments on human body *FIBER* **60(2)** 57–64

IOP Conf. Series: Materials Science and Engineering **684** (2019) 012011 doi:10.1088/1757-899X/684/1/012011

Study on physical properties of dispersed silica

E Tserenjav[1], V Jigmeddorj[2], B D Lygdenov[3], U Bayraa[4] and D D Tsevelmaa[5]

[1] Department of Chemical and Biological Engineering, School of Applied Science & Engineering, National University of Mongolia, 1 Ikh Surguuliin Gudamj, Ulaanbaatar 14201, Mongolia
[2] Department of Physics, School of Art & Sciences, National University of Mongolia, 1 Ikh Surguuliin Gudamj, Ulaanbaatar 14201, Mongolia
[3] Wuhan Textile University, 1 Sunshine Avenue, Wuhan 430073, China
[4] National University of Mongolia, 1 Ikh Surguuliin Gudamj, Ulaanbaatar 14201, Mongolia
[5] Mongolian University of Science and Technology, 34 Baga toiruu, 8th khoroo, Sukhbaatar district, Ulaanbaatar 14191, Mongolia

e-mail: erdenebatt@seas.num.edu.mn

Abstract. The paper deals with the physical properties of colloidal silica dispersed system such as distribution width (Q_x, %) and density (q_x, %) defined by photon cross correlation spectroscopy (PCCS) at an average size of 17.06 nm for D_{50} % and the area of specific surface of 349.96 m^2 cm^{-3} depending on molar (R_{m2}) optimal ratio of silica stone mineral to alkali ash powder. The change of crystal structure to amorphous state was determined by X-ray diffraction analysis. The results of their kinematic viscosity, liquid density, UV light adsorption inspections show possibility to produce the colloidal silica by using the wetting and ultrasonic vibration method for that fusible alloy.

1. Introduction

Silicon dioxide (SiO_2) nanoparticles aqueous dispersion is used in a wide range of environmentally friendly products in a large variety of industries: biomedicine, food and beverages field. Within this study dispersed silica particles are produced by the direct fusion of precisely determined ratios of soda ash (Na_2CO_3) and natural silica stone in electric muffle furnace at approximately 1100 °C and then it is not well dispersed typically after wetting [1–3].

But for most silica applications especially when being used in coatings to improve scratch resistance, the silica particles need to be smaller than 40 nm to fulfill this requirement and that particle agglomeration hinders each individual silica particle to interact with the surrounding media. So, the stability and agglomeration state of this aqueous dispersion nanofluids are the key parameters for their use [4–6].

Experimentally ultrasonic processing has been used to get more effective colloid size, stable and non-agglomerated dispersion system. The dispersed silica colloid particles were measured by using photon cross correlation spectrometer. The minimal particle size was reduced to 17.06 nanometers at D_{50} % of the test. The area of specific surface (S_v) for R_{m2} was obtained 349.9 m^2 cm^{-3} at molar optimal ratio of 2 for silica to sodium oxide weight.

2. Experimental procedure

2.1. Material and method

Dispersed silica nanoparticles were prepared by melting process of mixes by certain ratio of grinded powder of local silica stone from Tuv province, Sereglen soum located at 47°65'03.72"N, 107°25'22.47"E area with the utilization license N.21101 and sodium oxide as from sodium ash, CAS 497-19-8, Sigma Aldrich, Na_2CO_3. The physical and chemical properties were designed according to molecule ratio of acid to alkali as silica to sodium oxide weight which is shown in tables 1 and 2.

Table 1. Molar mixing designed ratio (R_m) of silica to sodium ash & oxide.

Sample no.	SiO_2	Na_2CO_3	Na_2O	$R_m=SiO_2/Na_2O$
1	10.1	8.1	3.4	3.0
2	10.1	12.2	5.1	2.0
3	6.0	14.5	6.0	1.0

Dispersed silica sols as yield content, density and formula for nanofluids etc. were prepared under their molar ratio (R_m) of the weight as shown in table 2.

Table 2. Physical and chemical characterization for obtained dispersed silica materials from Na_2SiO_3.

R_m	Yield (%)	Density (kg l^{-1})	Formula
3.0–3.2	25–40	1.26–1.45	$Na_2O \cdot xSiO_2 + H_2O$ ($3.0 < x < 3.2$)
2.0–2.6	35–46	1.38–1.53	$Na_2O \cdot xSiO_2 + H_2O$ ($2.0 < x < 2.6$)
1.0–1.6	35–55	1.38–1.70	$Na_2O \cdot xSiO_2 + H_2O$ ($1.0 < x < 1.6$)

Chemical composition of the silica stone used in the experiment was shown in the table 3 by comparing the test results of certified reference material (CRM.N40 testing) and its certified data of CRM.N40 according to the certificate. Mineral composition of the silica stone was studied by powder X-Ray Diffraction Analysis (XRDA) with the Siemens Diffractometer D5000 equipment, the results described in figure 1.

Table 3. Chemical composition of the silicastone.

Sample	SiO_2	Al_2O_3	TiO_2	Fe_{tot}	CaO	MgO	Na_2O	K_2O	MnO	P_2O_5	LOI
Silicastone	98.56	0.15	< 0.005	1.13	< 0.01	< 0.01	< 0.01	0.03	0.012	< 0.005	0.28
CRM.N40 test	87.98	6.15	0.057	0.27	0.08	0.16	0.04	4.51	0.008	0.076	0.57
CRM.N40 cert	88.20	6.18	0.058	0.261	0.11	0.15	–	4.23	–	0.077	–

The dispersed colloidal silica was prepared in the following steps [7]: mixing initial components, fusing them to temperature influence as approximately 1000 °C and wetting by distilled water to transform it into liquid state. The cleaning & non-agglomeration was done using centrifugal and ultrasonic method [8] with 800D Centrifuge equipment at 2000-2500 rpm speed and SK – 120 DTH Ultrasonic with 900-watt capacity for 15 min each sample. Value of pH for synthesized sol was defined with the laboratory PH5011 pH-meter. Liquid density of colloidal silica was determined by the aerometric method described in GOST 18481-81 [9] and kinematic viscosity was done by viscos meter described in GOST 10028-81 [10]. Absorptivity of UV light for dispersed nanoparticles in distilled water was determined by spectroscopic method using Shimadzu, UV-Vis Recording Spectrophotometer, UV-2401PC. The size and the value of average suspended sol, and the width of particle size distributions etc. were defined by photon cross correlation spectroscopic (PCCS) method using Shimadzu, STMPA

IOP Conf. Series: Materials Science and Engineering **684** (2019) 012011 doi:10.1088/1757-899X/684/1/012011

Nanophox as described in MNS:ISO 13321:2014 standard [11]. This equipment determines the particle size from few nanometers to value 10 mkm at the initial step of precipitation. Test results were calculated by using WINDOX 5 program.

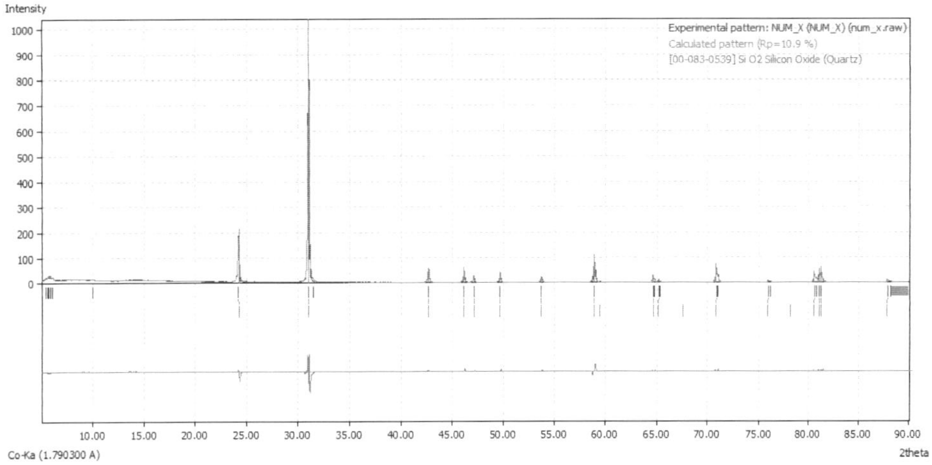

Figure 1. X-ray powder pattern of the silica stone.

3. Results and discussion

The chemical composition of silica stone was used in this study, as shown in table 3. X-ray diffraction analysis was performed for both untreated silica stone (figure 1) and the dispersed silica, either treated by different ratio for sodium ash at 1100 °C for 1 h (figure 2).

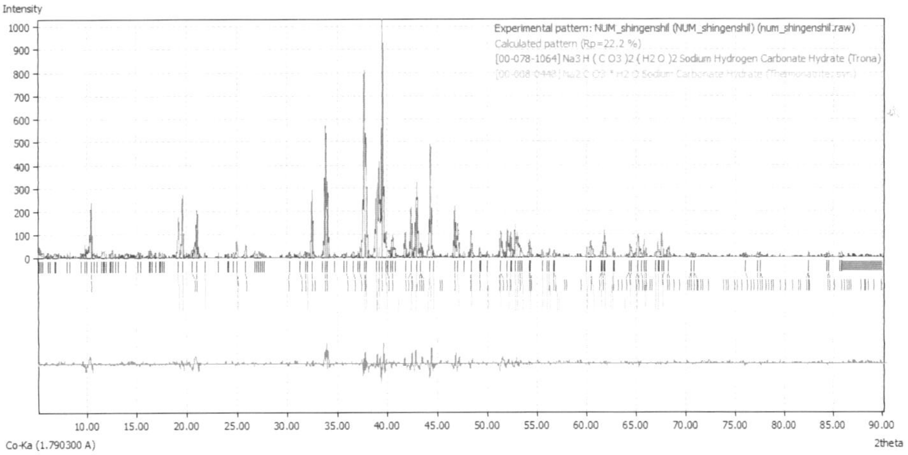

Figure 2. X-ray pattern of colloidal silica sol with sodium ash.

Measurement of all these properties provides information about the colloidal state of nanofluids. Hence the most important variable was the solid content. As it is shown in figure 1, X-ray diffraction intensive peaks corresponding to quartz crystal of silica disappeared and decreased their reflection intensity in figure 2, because of that at x-ray pattern with sodium carbonate phase after drying this low melting alloy was dispersed from the surface of colloidal amorphous silica phase at a higher temperature by distilled water (see figure 2).

By X-ray diffraction analysis the crystal structure of quartz crystal disappeared at $2\theta = 31.8°$, having confirmed its complete transition to amorphous state for this pyro chemically prepared colloidal silica particles. Behind it, pH value of the medium can be used as an important factor for the stability of system to control it and as a driving force in electric chemistry [12]. So, the value of pH and density (ρ, g ml^{-1}), kinematic viscosity (μ), absorbed spectroscopy investigations for UV light were measured on all synthesized colloidal silica samples and shown in table 4 and figure 3.

Table 4. Study on molar ratio dependence of pH, density, viscosity and UVabs for colloidal nanofluids.

R_m	pH	ρ (g ml^{-1})	μ (Pa s)	UVabs (nm)
1	13.53	1.08	226.62	303
2	12.94	1.09	307.85	361
3	12.3	1.035	576.08	333

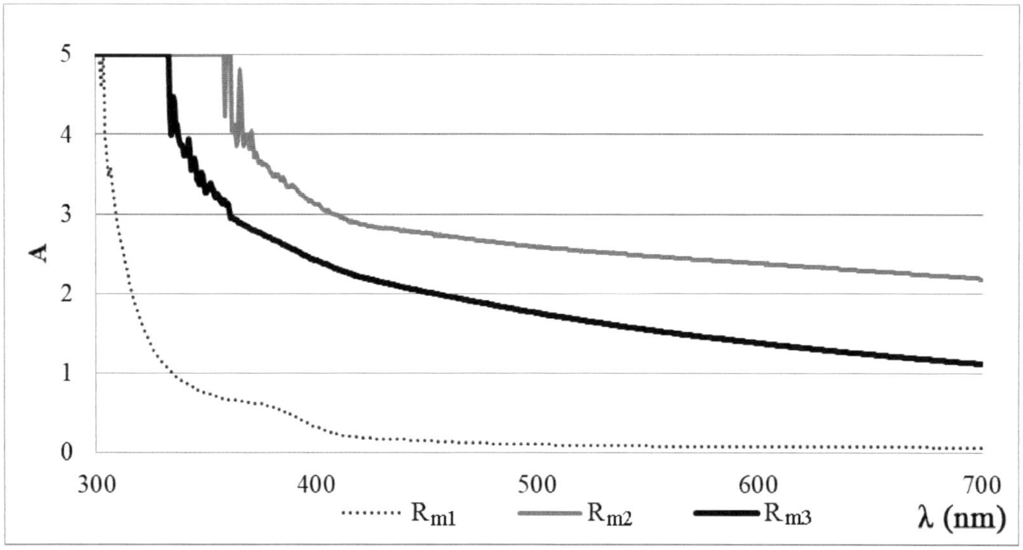

Figure 3. Absorbed UV spectrum of colloidal silica sol.

The band gap of electronic transition of vibrational fine structure for the absorptivity of UV-Vis spectra at the widest value of l = 361 nm area was observed for R_{m2} type soluble silica suspension. These results showed that dispersions can be used or diluted with suitable (compatible) solvents such as suspensions of nanoparticles in water. These dispersions of nanoparticles can sometimes settle while storage, in this case they can be mixed (shaken up) before use and that colloidal silica is used as a raw material for water glass production optimal at molar ratio R_{m2} for silica to alkali metal oxide. In addition, this dispersed phase could not dissolve in the dispersion medium of water with negative charges observed while their aging time and dispersion was controlled with the photon cross correlation spectroscopy to achieve ultrasonic treatment. Table 5 presents in detail the experimental results of the particles with the size of the dispersed silica sol depending on its distribution width (Q_x, %) and density of colloidal suspension (q_x, %), and coefficient of regression (r_2) for linear equation for all samples, this also shown in figure 4.

The mass median diameter as D_{50} for R_{m1} = 43.9 nm, R_{m2} = 17.1 nm, R_{m3} = 44.8 nm was shown in figure 4 and in table 5. Concerning the influence of the solid content, pH and salt concentration results declared that minimal average size of colloidal silica from the area of specific surface was 349.96 m^2 cm^{-3} at R_{m2}. So, to produce colloidal SiO_2 particles the molar optimal ratio was concluded to be 2 when using sodium carbonate to process the natural silica stone minerals.

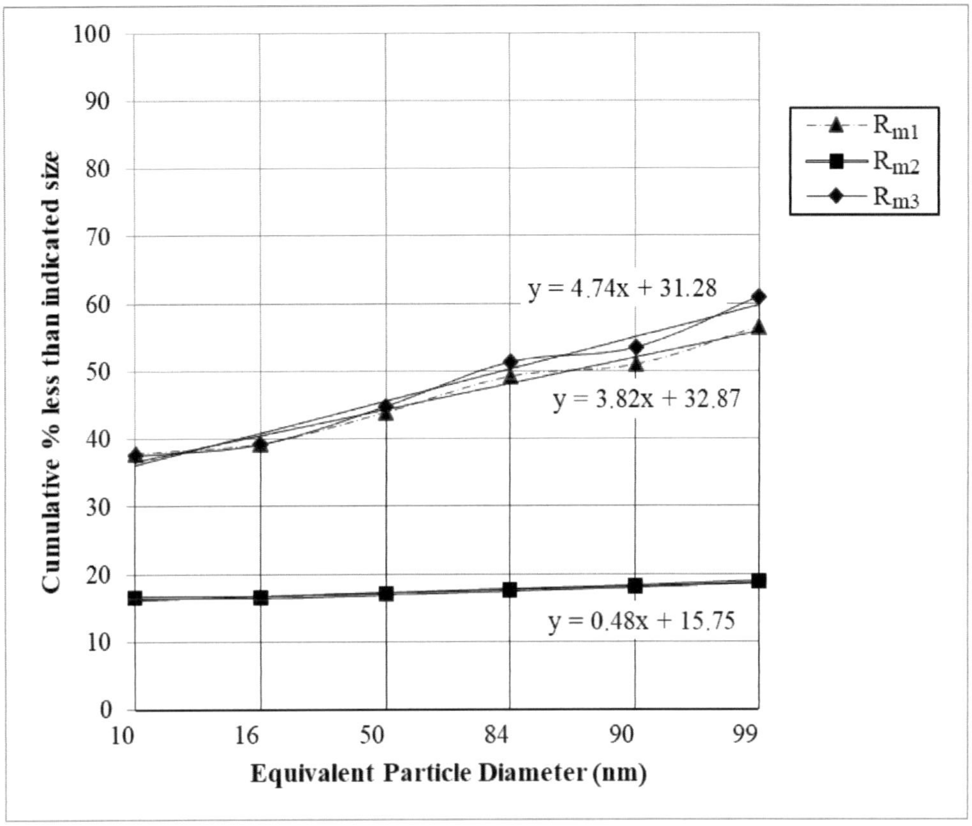

Figure 4. Colloidal silica particles distribution statistical curve by cumulative percent.

Table 5. Average size (X_{nm}, %), density (q_x), area (Q_x, %) and regression (r_2) coefficient of particle size distribution width for nano dispersed colloidal silica.

R_m	D_{nm}, %						SV (m^2 cm^{-3})	Slope σ_g = x84 x50^{-1}	$\alpha = \sigma_g$ x50^{-1}	r_2	SMD (nm)	VMD (nm)
	10	16	50	84	90	99						
1	37.8	39.2	43.9	49.2	50.9	56.5	137.58	1.12	0.03	0.98	43.61	44.15
2	16.5	16.6	17.1	17.7	18.1	18.8	349.96	1.04	0.06	0.97	17.14	17.16
3	37.6	39.2	44.8	53.3	53.4	60.9	135.05	1.19	0.03	0.96	44.43	45.22

4. Conclusion

Colloidal silica was produced by pyro chemical of silica stone, being an abundant mineral in Mongolia with molar different ratios of sodium ash at 1100 °C. The synthesis parameters, such as molar ratio of mixing, fluid density and size of colloid particles were optimized.

X-ray powder diffraction of colloidal silica disappeared at the crystal peak of quartz peak at $2\theta = 31.8°$ and UV-vis. Spectrum. The band gap vibration of electronic transition was observed at the widest value of l = 361 nm area for R_{m2} type soluble silica suspension.

Colloid silica nanoparticles were confirmed at the optimized parameters of R_{m2} by PCCS inspection with the area of specific surface of up to 349.96 m^2 cm^{-3}. The average size of the colloidal silica particles was 17.6 nm at D_{50} % value, coefficient of regression for linear equation being 0.97 and with 1.04 slope.

These findings seem to be extremely important for the industry to produce colloidal SiO_2 particles by using silica stone and sodium ash minerals in practice in areas given.

Acknowledgments
The authors kindly appreciate the support of National University of Mongolia for their funding advanced technology promotion of the P2016-1218 project "Getting the method of soluble silica for leather processing industry from abundant silicate minerals of Mongolia" and the colleagues of Laboratory for New Materials at Dept. of Chemistry & Biological Engineering, School of Engineering & Applied Sciences.

References

[1] Neville F, Mohd Zin A, Jameson G J and Wanless E J 2012 Preparation and characterization of colloidal silica particles under mild conditions *J. Chem. Educ.* **89(7)** 940–2
[2] Kobayashi Y et al. 2016 Fabrication of quantum dot/silica core–shell particles immobilizing Au nanoparticles and their dual imaging functions *Appl. Nanosci.* **6(3)** 301–7
[3] Bitter J L, Duncan G A, Beltran-Villegas D J, Fairbrother D H and Bevan M A 2013 Anomalous silica colloid stability and gel layer mediated interactions *Langmuir* **29(28)** 8835–44
[4] Buckley A M and Greenblatt M 1994 The sol-gel preparation of silica gels *J. Chem. Educ.* **71(7)** 599
[5] Hoshikawa Y, Yabe H, Nomura A, Yamaki T, Shimojima A and Okubo T 2010 Mesoporous silica nanoparticles with remarkable stability and dispersibility for antireflective coatings *Chem. Mater.* **22(1)** 12–4
[6] Sekhar P K and Bhansali S 2016 Growth of silica nanowires *Encyclopedia of Nanotechnology* (Dordrecht: Springer Netherlands) 1404–27
[7] Mondragon R, Julia J E, Barba A and Jarque J C 2012 Characterization of silica–water nanofluids dispersed with an ultrasound probe: a study of their physical properties and stability *Powder Technol.* **224** 138–46
[8] Serrano-Ruz J A et al. 2018 Synthesis of colloidal silicon and germanium nanoparticles by laser ablation of solid Si and Ge targets in ethanol *Mater. Res. Express* **5(1)** 015038
[9] 2007 *GOST 18481-81 Glass Hydrometers and Cylinders. General Specifications* (Moscow: Standartinform)
[10] 2005 *GOST 10028-81 Glass Capillary Viscosimeters. Specifications* (Moscow: Standartinform)
[11] 2017 *Mongolian Standards Catalogue. List of Standards Classified by Subject* p 131
[12] Tang B, Wu C, Qiu M, Zhang X and Zhang S 2014 PEG/SiO2–Al2O3 hybrid form-stable phase change materials with enhanced thermal conductivity *Mater. Chem. Phys.* **144(1–2)** 162–7

IOP Conf. Series: Materials Science and Engineering **684** (2019) 012012 doi:10.1088/1757-899X/684/1/012012

Synthesis and antimicrobical activity of composite oxides nanoparticles based on ZnO

A M Kondranova, A S Lozhkomoev, S O Kazantsev and A V Pervikov

Institute of Strength Physics and Materials Science of Siberian Branch of Russian Academy of Sciences, 2/4 Akademicheskii Avenue, Tomsk 634055, Russia

e-mail: amk@ispms.tsc.ru

Abstract. It should be noted that zinc oxide nanoparticles have increased antibacterial activity. ZnO/Me composite nanoparticles have a more pronounced antimicrobial effect in comparison with monometallic nanoparticles obtained by similar conditions. In present work ZnO/Ag and ZnO/CuO composite nanoparticles were obtained by the electrical explosion of Zn and Cu or Zn and Ag wires in oxygen containing atmosphere. Composite nanoparticles were characterized by transmission electron microscopy, XRD analysis and micro electrophoresis. Antibacterial activity of composite nanoparticles against Gram-positive *E.coli* and Gram-negative MRSA was studied. The as-synthesized nanoparticles demonstrated excellent bactericidal effect due to their unique properties including not only components additive effect but also synergetic one.

1. Introduction

Nanoparticle metal oxides represent a new class of materials that are increasingly being developed for use in research and medical applications. These metal oxides are interesting not only for a wide range of physical and chemical properties, but also for antibacterial activity [1]. And the problem of formation of stable strains of microorganisms contributes to the development of directions related to the development of nanomaterials, which can become a new alternative to the already known antibiotics [2]. Throughout the world, there is one complex problem that modern bacterial-infection therapy faces: the emergence of drug-resistant microorganisms, such as methicillin-resistant *Staphylococcus aureus* (MRSA) [3]. And according to preliminary experimental data, zinc oxide-based composite nanoparticles influence the clinical strain of MRSA. This makes it possible to develop promising areas of research based on zinc oxide composite materials. And the development of these new composite nanoparticles will allow to obtain materials with increased functional properties and antibacterial activity. Thus, for example, nanocomposites based on metal oxides are used to create new effective antimicrobial agents [4, 5]. Zinc oxide (ZnO) is an important metal oxide used for various biological applications [6, 7]. Such composite nanomaterials based on zinc oxide with antimicrobial activity are of great interest for various biomedical applications.

2. Materials and methods

Composite nanoparticles $(Zn_xMe_{1-x})O$, where Me – Cu, Ag were obtained by a electric explosion of two intertwined wires in a buffer gas containing argon and oxygen. The oxygen content was 20 vol. %. The buffer gas pressure for all samples was $3 \cdot 10^5$ Pa.

2.1. The morphology of nanoparticles
Composite nanoparticles morphology and particle size distribution were studied by the transmission electron microscopy (TEM) method using the electron microscope JEM-2100 (JEOL, Japan) with the integrated energy-dispersive analysis system X-Max (Oxford Instruments, GB). The phase composition of the samples was studied by X-ray analysis (XRD) on CuKα radiation in the scanning mode in the range of 2θ angles from ~25 to ~75°, with a pitch of 0.02 using a XRD Shimadzu 6000 diffractometer (Shimadzu, Japan). The zeta potential of nanoparticles was measured in deionized water at 25 °C and pH 7.2 using the Zetasizer Nano ZSP (Malvern Instruments Ltd, GB).

2.2. Determination of the minimum inhibitory concentration
The nanoparticles antibacterial activity was determined using Gram-positive clinical strain of methicillin-resistant Staphylococcus aureus MRSA ATCC 43300 and a Gram-negative Escherichia coli ATCC 25922. The minimum inhibitory concentrations (MIC) values were determined by 96-well plate microdilution method. In a typical procedure, 100 μl of Mueller Hinton broth, 20 μl of bacterial suspension (108 CFU/ml) and 50 μl of aqueous dispersions of nanoparticles were added to each well.

The plates were then incubated at 37 °C for 24 hours. After incubation, optical density (OD) was measured at 620 nm using the Multiskan FC microplate spectrophotometer (Thermo Fisher Scientific, USA).

To determine the bacterial growth dynamics in the phosphate buffered saline (PBS) in the presence of nanoparticles a suspension method was used. The 300 μg sample weights were added to 30 ml PBS. After that, 300 μl bacterial suspensions with the concentration of 10^7 CFU/ml was added and incubated at room temperature under stirring on a PE-6600 shaker (Ecroskhim, Russia). The 50 aliquots μl were sampled from each container after 1, 3, 6, 9 and 24 hours and distributed on Petri dishes with Mueller Hinton agar. The number of colony-forming units (CFU) was calculated after 24 hours incubation at 37 °C. Microbial suspensions without nanoparticles in sterile PBS were used as controls.

3. Results and discussion

3.1. Production of nanoparticles
As a result of joint electrical explosion of two intertwined Zn/Ag, Zn/Cu wires (EEIW) in the oxygen-containing atmosphere, nanoparticles and their agglomerates of various size are formed. For a sample $(Zn_{0.55}Cu_{0.45})O$ the maximum quantity of particles has the number average particle size of particles in the range of 70-100 nm, for a sample $(Zn_{0.55}Ag_{0.45})O$ more than 40 % of all particles have the size in the range of 20-50 nm. At electrical explosion of couple of Zn/Cu metals particles, spherical and close to spherical shape, are formed. At the same time copper and zinc on particles are distributed not evenly, division of components (figure 1(a)) is observed. The sample received at dispersion of Zn and Ag (figure 1(b)) delays, represents particles of irregular shape. At the same time the form Janus particles, where one of parts is enriched with Zn, other Ag is characteristic of many particles and there is a clear boundary of division of components.

3.2. Analysis of composite nanoparticles
According to the X-ray analysis data, at joint electro-pulse dispersion of Zn/Cu metals in an oxygen-containing atmosphere (figure 2(a)) a mixture of ZnO phases (PDF card 01-075-0576), CuO (PDF card 00-002-1040) and Cu_2O (PDF card 01-078-5772) is formed. The main peaks on the diffractogram of the sample obtained at dispersion of Zn/Ag pair (figure 2(b)), peaks correspond to ZnO (PDF card 01-075-0576) and Ag (PDF card 01-071-3762). Thus, under the experimental conditions silver does not react with oxygen.

All particles have a positive zeta potential in water, measured at 25 °C and at physiological pH values. At the same time, the zeta potential of ZnO–CuO nanoparticles at pH 7 is approximately +40 mV, ZnO–Ag(O) nanoparticles - +28 mV. The point of zero charge (IEP) of nanoparticles is in the pH 9-9.7 range.

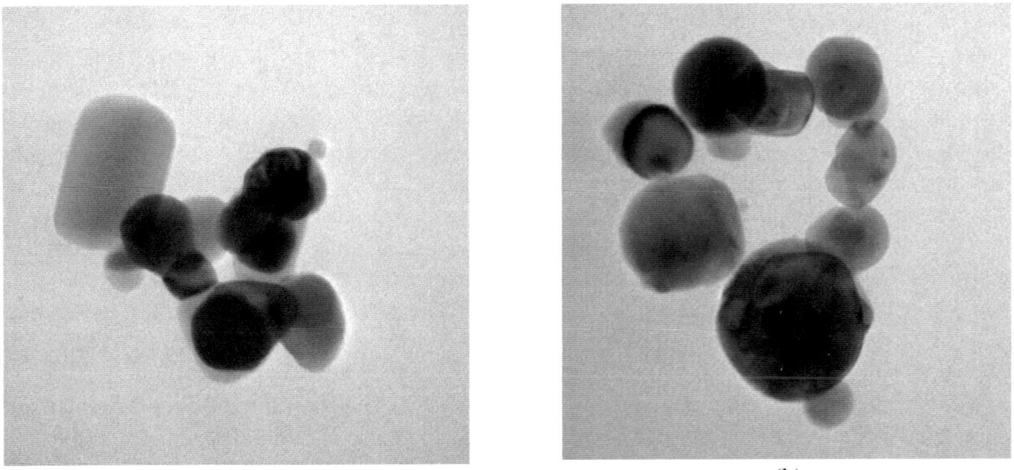

(a) (b)

Figure 1. TEM images of (a) $(Zn_{0,55}Cu_{0,45})O$, (b) $(Zn_{0,55}Ag_{0,45})O$ nanoparticles.

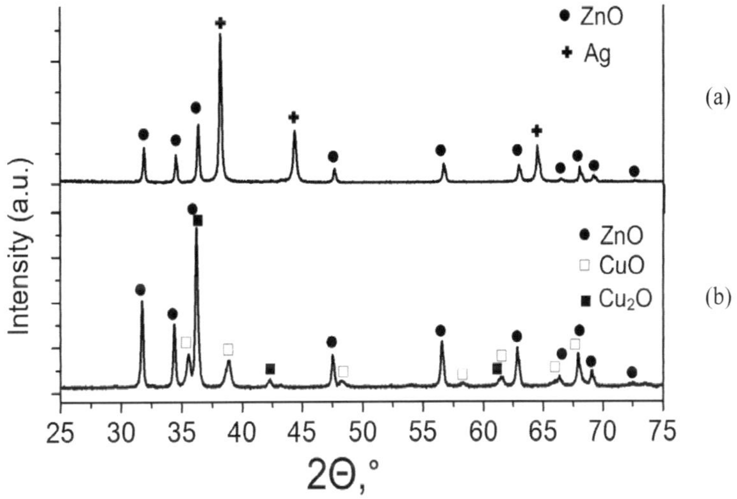

Figure 2. XRD pattern of (a) $(Zn_{0,55}Cu_{0,45})O$, (b) (Z $(Zn_{0,55}Ag_{0,45})O$.

At electrical explosion in the oxygen-containing atmosphere of Zn, Cu, Ag wires nanoparticles are formed, having the following characteristics: at dispersion of Zn the faceted particles of ZnO and their agglomerates are formed, the majority of which have the average particle size in the range of 20-70 nm). More than 80 % of the particles are in the 20-100 nm range. The electrical rupture of a Cu conductor results in the formation of spherical nanoparticles, 45 % of which are in the 70-100 nm range. The particles contain CuO and Cu_2O, as well as a low intensity reflex characteristic of Cu. Silver under these conditions does not react with oxygen and as a result of electrical explosion of Ag wire spherical silver nanoparticles are formed, 77 % of which are in the range of 0.2-0.7 nm.

All particles have a positive zeta potential in water, measured at 25 °C and at physiological pH values. The zeta potential of $(Zn_{0,55}Cu_{0,45})O$ nanoparticles at pH 7 is approximately +40 mV, $(Zn_{0,55}Ag_{0,45})O$ nanoparticles - +28 mV. The point of zero charge (IEP) of nanoparticles is in the pH 9-9,7 range. Thus, under conditions of combined electro-pulse dispersion of zinc and the second metal in an oxygen-containing atmosphere, composite nanoparticles consisting of zinc oxide, second metal oxide, or non-responsive metal with oxygen can be obtained.

3.3. Antibacterial activity

In the study of antimicrobial activity of composite nanoparticles, it was found that the nanoparticles under study most effectively inhibit the growth of MRSA (clinical strain) ATCC 43300 compared with particles that were obtained by dispersing single conductors in the same conditions. The minimum inhibitory concentration for $(Zn_xMe_{1-x})O$ nanoparticles is several times lower than for Me(O) nanoparticles (figure 3). Preliminary growth analysis data show that ZnO-based nanoparticles have a much higher antibacterial effect on *Staphylococcus aureus*. But composite nanoparticles also affect *E.coli* ATCC 25922, but are not as pronounced. The data obtained suggest that nanoparticles containing ZnO can potentially be used as a new promising bacteriostatic agent and used in the development of derivative agents to control the spread and infection of various strains of bacteria. At the same time, the results of the study of the dynamics of microorganism reduction during the incubation of MRSA with composite nanoparticles $(Zn_{0,55}Cu_{0,45})O$ and $(Zn_{0,55}Ag_{0,45})O$ in PBS show that after 1 hour of incubation, the complete suppression of bacterial activity occurs (figure 3(a), (i)). And single nanoparticles of metals (ZnO, AgO, CuO) showed bactericidal action only after 3 hours of incubation (figure 3(c), (e), (g)). And on *E.coli*, complete suppression of microorganisms occurs after 3 hours of incubation with nanoparticles Ag(O), $(Zn_{0,55}Ag_{0,45})O$, CuO (figure 3(d), (f), (j)) and after 1 hour at $(Zn_{0,55}Cu_{0,45})O$ (figure 3(b)).

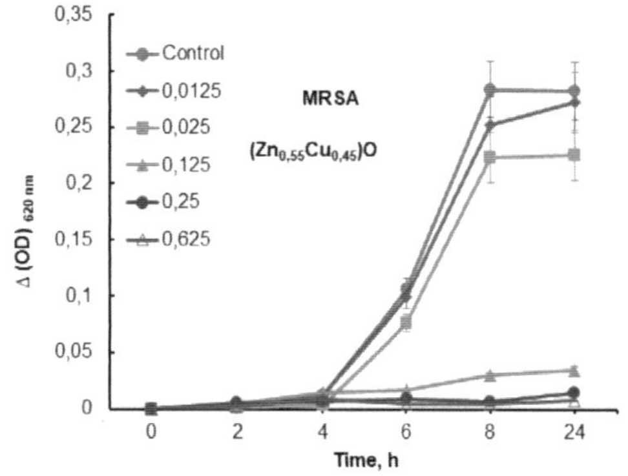

(a)

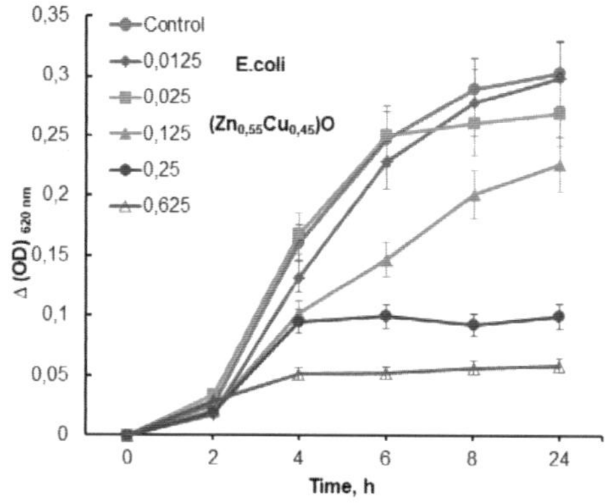

(b)

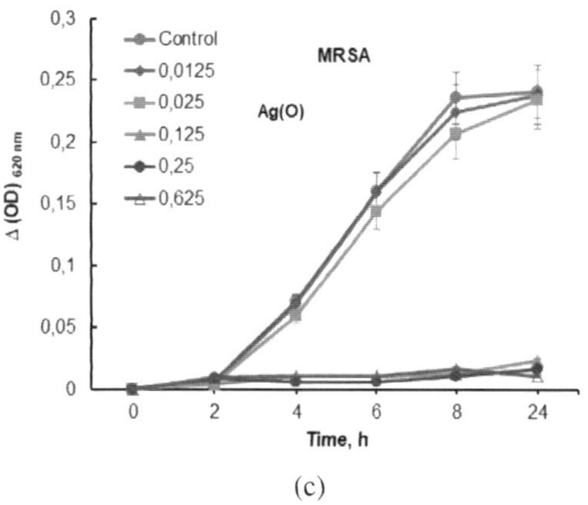

(c)

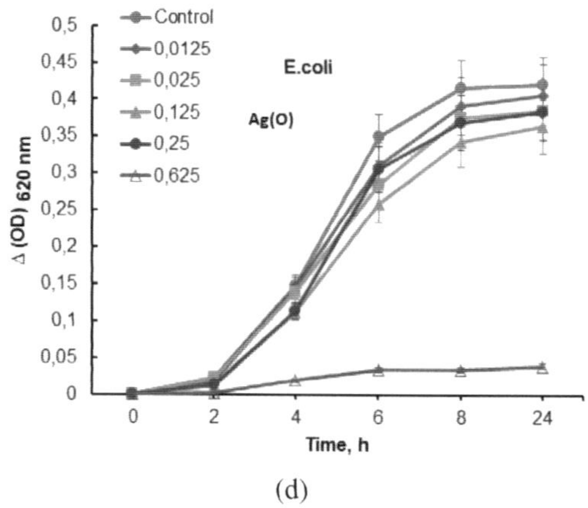

(d)

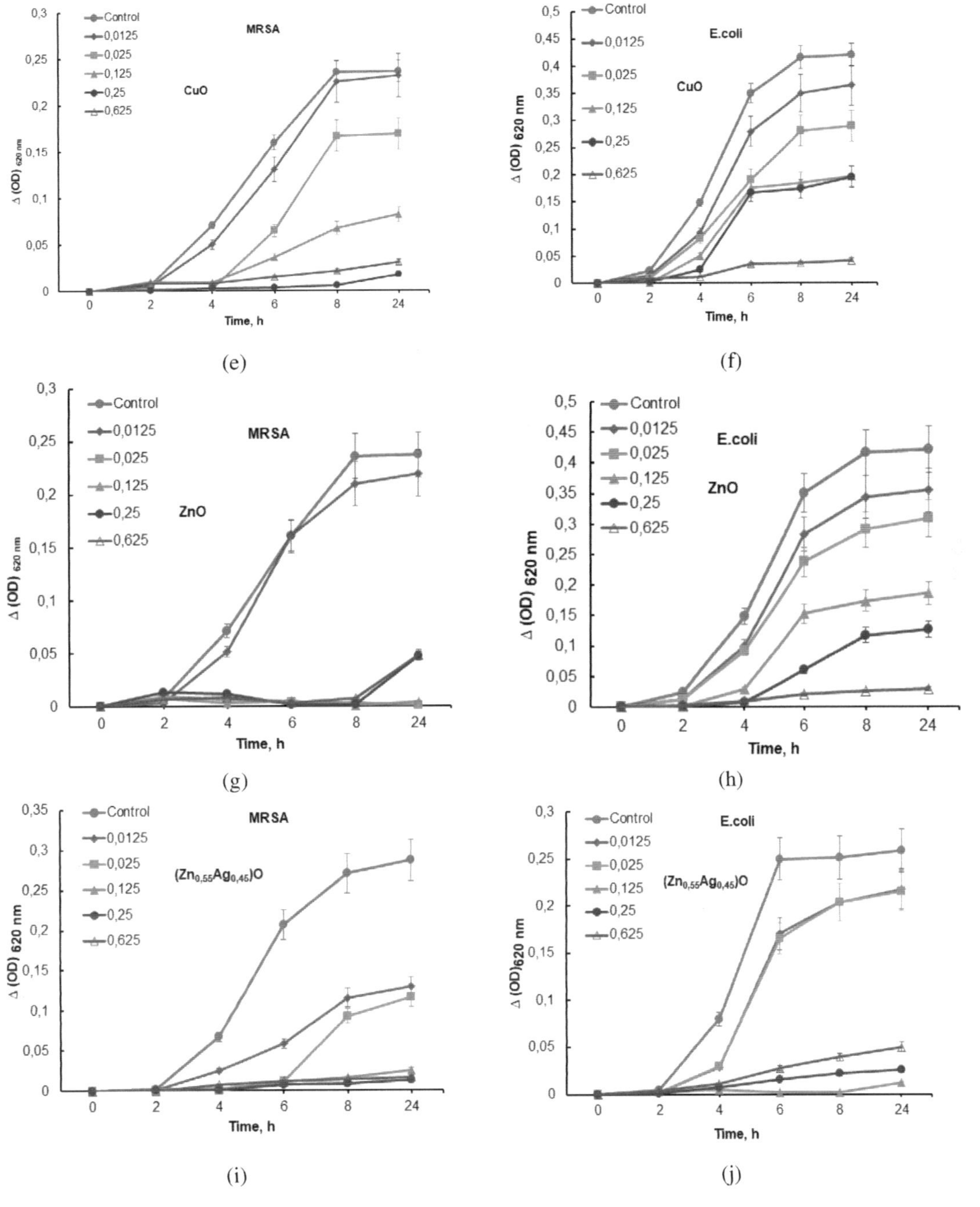

Figure 3. MIC of $(Zn_xMe_{1-x})O$ nanoparticles.

4. Conclusions

Composite nanoparticles the received electrical explosion of two intertwined Zn/Ag, Zn/Cu wires in the oxygen-containing atmosphere can turn out in the form of nanoparticles of various structure, such as mix of oxides or as in case of joint electric explosion of Zn and Cu of wires, mix of oxides and spinels. The synthesized composite nanoparticles have the greatest antimicrobial activity. They show bactericidal actions on Gram-positive MRSA and on Gram-negative *E.coli*, than the nanoparticles received in similar conditions at electro pulse dispersion of single wires of the corresponding metals, including silver in the oxygen-containing atmosphere. The presented results, at more in-depth study biological action of the received nanoparticles, can find application for development of new antibacterial agents and the strategy of fight against pathogenic microorganisms.

Acknowledgments
The work was financially supported by the Russian Science Foundation, grant No. 17-79-20382.

References

[1] Jones N, Ray B, Ranjit K T and Manna A C 2008 *FEMS Microbiol. Lett.* **279(1)** 71–6
[2] Alswat A A, Ahmad M B , Saleh T A, Hussein M Z B and Ibrahim N A 2016 *Mater. Sci. Eng.* **68** 505–11
[3] Koch G, Yepes A, Förstner K U, Wermser C, Stengel S T, Modamio J and Lopez D 2014 *Cell* **158(5)** 1060–71
[4] Kaushik M, Niranjan R, Thangam R, Madhan B, Pandiyarasan V, Ramachandran C and Venkatasubbu G D 2019 *Appl. Surf. Sci.* **479** 1169–77
[5] Azam A, Ahmed A S, Oves M, Khan M S, Habib S S and Memic A 2012 *Int . J. Nanomedicine* **7** 6003
[6] Arias C A and Murray B E 2009 *N. Engl. Med.* **360(5)** 439–43
[7] Balouiri M, Sadiki M and Ibnsouda S K 2016 *J. P. A.* **6(2)** 71–9

IOP Conf. Series: Materials Science and Engineering **684** (2019) 012013 doi:10.1088/1757-899X/684/1/012013

Mathematical modelling of the process of melting alloys of the Si-Cr-Ni-Fe system in the iron-carbon melt

O V Zayakin[1], D S Renev[1], V I Zhuchkov[1] and E Yu Lozovaya[2]

[1] The federal state budgetary scientific institution "Institute of Metallurgy of Ural branch of the Russian Academy of Sciences", 101 Amundsen Street, Yekaterinburg, 620016, Russia

[2] Ural Federal University named after the first President of Russia B. N. Yeltsin, 19 Mir Street, Yekaterinburg, 620002, Russia

e-mail: zferro@mail.ru

Abstract. The process of melting of the Si-Cr-Ni-Fe ferroalloys in the iron-carbon melt under static conditions was studied using the method of mathematical modelling. It was found that the melting process takes place in three stages. The effect of silicon concentration and the initial size of a piece of alloy on the duration of each stage of alloy melting is revealed. It was shown that an increase in the silicon content from 6 to 40 % in the complex ferroalloy of the 50 mm fraction containing ~ 27 % Cr and ~ 12 % Ni leads to a decrease in the total melting time of the alloys from 67 to 30 seconds at the iron-carbon melt temperature of 1600 °C.

1. Introduction

In recent years the production of corrosion-resistant chromium-nickel steels has significantly increased in the world [1, 2]. In Russia electrolytic nickel is mainly used for smelting of such steels, the high price of which doesn't promote the development of steelmaking [3]. At the same time there is a shortage of relatively cheap complex nickel ferroalloys in Russia [4, 5]. The development and application of this type of ferroalloys requires awareness of their physicochemical properties [6–8]. One of the most important properties of the alloy intended for alloying steel is the melting time and dissolution in the iron-carbon melt that determines both the recovery and distribution of the principal alloying elements of the ferroalloy in the metal volume [9]. The melting mechanism of alloys is a fundamental factor determining the time of its melting in an iron-carbon melt. Extensive research has been carried out in this area by both domestic [10, 11] and foreign [12–14] metallurgists.

2. Materials and methods

In this work to study the melting process of complex nickel-containing alloys in an iron-carbon melt a mathematical model for calculating the melting time developed by scientists of the Ural Federal University and the IMET Ural Branch of the Russian Academy of Sciences was used. The model includes hydrodynamic and thermal parts. The thermal part of the model is based on the classification of ferroalloys depending on the ratio of their liquidus temperatures (T_l), bath temperatures (T_b) and crystallization temperatures of the iron-carbon melt (T_{cm}). Thus the ferroalloys are divided into low melting ($T_l \le T_{cm}$), high melting ($T_{cm} < T_l < T_b$) and ultra-high melting ($T_l \ge T_b$).

The samples of complex alloys of the Si-Cr-Ni-Fe system, corresponding in composition to ferroalloys that may be smelt from domestic mineral raw materials, were selected for the study [15, 16] (table 1).

Table 1. Chemical composition of ferroalloys.

	Content[a] (%)		
No.	Si	Ni	Cr
1	5.8	12.2	27.4
2	13.0	11.9	28.8
3	20.4	12.1	26.3
4	40.1	11.7	25.5

[a] ~ 0.2 % C, Fe – the rest.

3. Results and discussion

The simulation results are presented in figures 1, 2.

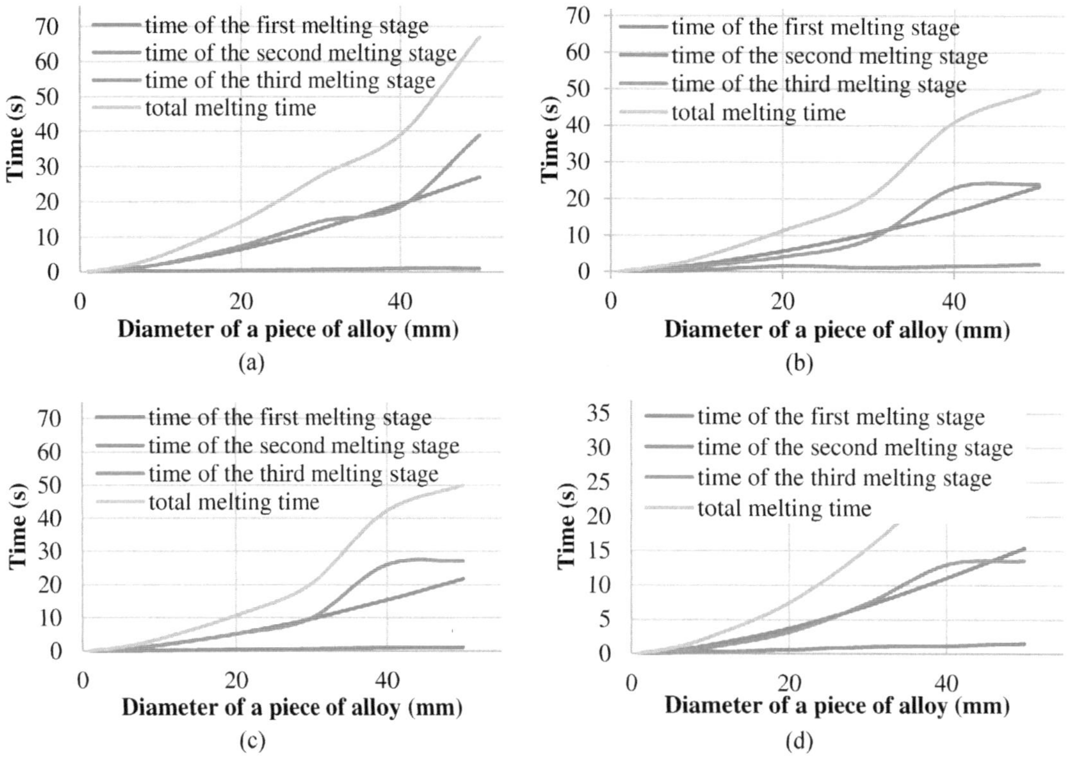

Figure 1. Dependence of the melting time of nickel-containing ferroalloys on the fractional composition at T_b = 1600 °C by melting stage: (a) – alloy 1; (b) – alloy 2; (c) – alloy 3; (d) – alloy 4.

In the calculations it was assumed that $T_{cm} = 1530\,°C$, the initial temperature of the ferroalloy $T_0 = 25\,°C$ and the temperature of the bath of the iron-carbon melt is 1600 °C. The physicochemical properties of the ferroalloys were taken from the calculated [17] and previously obtained experimental data [18]. Under real conditions the initial temperature of a piece of ferroalloy is always lower than the crystallization temperature of steel, therefore at first a crust of solid steel is formed on its surface. The thermal part of the model includes Fourier differential heat condition equation for each of the phases (crust, solid core, liquid layer of the ferroalloy) and boundary conditions determined by convective heat transfer and the melting process.

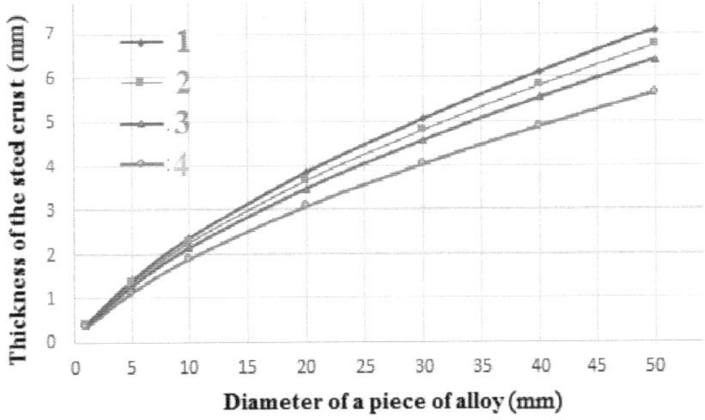

Figure 2. Dependence of the thickness of the solidified steel crust on the fractional composition of the ferroalloy at T_b = 1600 °C (the numbers of the curves correspond to the numbers of alloys in the table 1).

All considered alloys belong to the group of low melting ferroalloys in which $T_l \le T_{cm}$. The process of their melting in the iron-carbon melt takes place in three stages. During the first stage the ferroalloy warms up, and when the surface temperature reaches T it begins to melt. Pieces of ferroalloy in this case are in the shell of solid steel, the thickness of which increases at the beginning of the stage and decreases towards the end. In the second stage the ferroalloy melts under the crust of solid steel due to the heat coming from the iron-carbon melt. It ends with the complete melting of the solid shell and the contact of the not melted but rather heated part of the piece of ferroalloy with liquid steel. The longer the first stage lasts the more intensely the crust of the solidified steel warms up and the bigger part of it melts thereby reducing the duration of the second stage. For all studied alloys the second stage is the shortest (no more than 5 % of the total melting time for a fraction of 50 mm). In the third stage a solid piece of ferroalloy melts directly contacting with liquid steel.

The influence of the initial size of the piece is unambiguous. With diameter decreasing, the total melting time is reducing (figure 1). With a decrease in ferroalloy fineness by a factor of 50, the total melting time decreases by a factor of 500–600.

An increase in the Si content from 6 to 40 % in the considered alloys leads to a decrease in the total melting time from 67 to 30 s (figure 1) and to a reduction in the thickness of the solidified steel crust from 7.1 to 5.6 mm (figure 2) with an initial ferroalloy diameter of 50 mm. This is due to the dependence of the thermophysical characteristics of the ferroalloys under study on the silicon concentration. In general, alloys of the Si-Cr-Ni-Fe system are characterized by a much faster melting process in steel compared to standard ferrochrome and ferronickel [9].

4. Conclusions

The process of melting ferroalloys of the Si-Cr-Ni-Fe system in an iron-carbon melt under static conditions was studied by mathematical modelling. It was revealed that the melting process proceeds in three stages. It was shown that an increase in the silicon content from 6 to 40 % in the complex ferroalloy of the 50 mm fraction containing ~ 27 % Cr and ~ 12 % Ni leads to a decrease in the total melting time of alloys from 67 to 30 sec at the temperature of iron-carbon melt of 1600 ºC. In general, alloys of the Si-Cr-Ni-Fe system are characterized by a much faster melting process in steel compared to standard ferrochrome and ferronickel.

Acknowledgments

This work was carried out according to the State Assignment of the IMET UB RAS and according to the RFBR project No. 19-03-00451.

IOP Conf. Series: Materials Science and Engineering **684** (2019) 012013 doi:10.1088/1757-899X/684/1/012013

References

[1] Pariser H H, Backeberg N R, Masson O C M and Bedder J C M 2018 *J. S. Afr. Inst. Min. Met.* **118(6)** 563–8

[2] Lukin A S, Komolova O A and Grigorovich K V 2019 *Proc. Conf. Int. Part. and Conf. Young Scient. Fundamental Research and Applied Development of Processes for Processing and Disposal of Technogenic Formations TECHNOGEN-2019* (Yekaterinburg: Ural Branch of the Russian Academy of Sciences) pp 99–102

[3] Zhuchkov V I, Leontiev L I, Selivanov E N, Zayakin O V and Babenko A A 2014 *Mat. Int. Scien.-Pract. Conf. Modern Trends in the Theory and Practice of the Extraction and Processing of Mineral and Technogenic Raw Materials* vol 2 (Yekaterinburg: UMTS UPI) pp 210–5

[4] Zayakin O V 2002 *Development of a Rational Composition of the Technology for the Production of Nickel-Containing Ferroalloys from Poor Oxidized Nickel Ores: Dis. Cand. Tech. Sciences* (Yekaterinburg)

[5] Khalezov B D, Zayakin O V, Gavrilov A S and Zhuchkov V I 2017 Hydro-, pyrometallurgical method for producing alloys of the Fe-Ni-Cr-Mn-Si system *Butlerov Communications* **52(10)** 111–7

[6] Zhuchkov V I, Zayakin O V and Maltsev Yu B 2001 *J. Melts* **1** 7–9

[7] Spanov S S, Zhunusov A K and Tolymbekova L B 2017 *Metallurgist* **60(11–12)** 1149–54

[8] Andreev N A, Zhuchkov V I and Zayakin O V 2013 *Rus. Met. (Metals)* **2013(6)** 418–9

[9] Zayakin O V, Zhuchkov V I and Lozovaya E Y 2007 *Steel Transl.* **37(5)** 416–8

[10] Zhuchkov V I, Zayakin O V, Leontiev L I, Sychev A V and Kel I N 2017 *Steel Transl.* **47(5)** 291–5

[11] Shurygin P M and Shantarin V D 1964 *Proc. USSR Academy of Sciences: Metal. Min.* **2** 38–40

[12] Agryropoulos S A 1984 *Iron Steelmaker* **11** 48–57

[13] Seaton C E, Rodriguez A A, González M and Manrique M 1983 *Transaction ISIJ* **23(1)** 14–22

[14] Agryropoulos S A 1981 Heat and mass transfer in metallurgical system *Proc. of the Int. Centre for Heat and Mass Transfer* vol 11, ed D Brain Spalding and N Afgan (Washington: Hemisphere Publishing Corporation) pp 20–8

[15] Zhuchkov V I, Zayakin O V and Zhdanov A V 2010 *Proc. 12th Int. Ferroalloys Cong. (Helsinki)* vol 1 (Helsinki: Outotec Oyj) pp 311–5

[16] Zhuchkov V I, Zayakin O V, Leontiev L I, Esenzhulov A B and Ostrovskii Ya I 2008 *Russ. Metall.* **8** 709–12

[17] Zinoviev V E 1989 *Thermophysical Properties of Metals at High Temperatures* (Moscow: Metallurgy) p 384

[18] Zayakin O V and Renev D S 2019 *KnE Materials Science* **(5)1** 297–303

IOP Conf. Series: Materials Science and Engineering **684** (2019) 012014 doi:10.1088/1757-899X/684/1/012014

Analysis of the possibility of using Cu-Au nanoclusters as bits in phase charge memory devices

D A Bashkova, Yu Ya Gafner, I S Zamulin and M F Kuznetsov

Katanov Khakas State University, 90 Lenin Street, Abakan, 655017, Russia

e-mail: bashkova.daria@yandex.ru

Abstract. This work analyses the potential of using copper-gold nanoalloy as a material for phase charge memory cells. For this purpose, smooth cooling of the melt of Cu-Au nanoclusters of 2.0 – 8.0 nm diameter was simulated with the molecular dynamics method, based on the modified tight-binding potential (TB-SMA). Particles of different chemical compositions were investigated. The gold content in the alloy varied from 10 to 90 percent. The authors also explored the influence of the cooling intensity on the character of the formed substructures. To reach the goal, the values of the heat removal rate were varied. As a result, nanoclusters of compositions $Cu_{90}Au_{10}$ and $Cu_{10}Au_{90}$ were found in a range of particles. Some parameters were selected for them, at which the stable development of amorphous and crystalline structures is possible.

1. Introduction

To date, the most common type of non-volatile devices is a planar flash memory built on transistors with a floating gate. However, the scaling of such cells causes some problems. So, with a decrease in the size of elements below 15 nm, leakage currents increase, which leads to the degradation of components and the loss of information. The diminution in the number of electrons stored in the cells increases the probability of stochastic effects, for example, crosstalk from neighboring elements. As a result, the device reliability deteriorates significantly. Although some manufacturers make planar flash memory using ten nanometer technologies, it is obvious that this type of storage devices has reached the limits of its technological development. Therefore, the problem of creating alternative non-volatile memory devices is still relevant.

Experts regard phase charge memory (PCM) as one of the main candidates for the role of universal memory. Its operation principle is based on the phenomenon of a sharp increase in the specific conductivity of a material during the transition from the amorphous state to the crystalline one. These storage devices have a few affirmative characteristics, such as high reading and recording speeds, significant cell scalability potential and durable service life (up to 10^{12} rewriting cycles).

The elementary PCM cell is a system of the control transistor and two electrodes with a layer of phase charge material (1T1R type) between them. In the process of information recording, the voltage is applied to the electrodes in the amount sufficient for the electronic breakdown of the working layer and the formation of a conducting channel. The warmth dissipated in this area heats the active area to the temperature above crystallization. The PCM material transforms into a low resistance state (SET) after the pulse termination, i.e. it crystallizes. For data cancellation it is necessary to energize a short current pulse of higher amplitude. It heats the work area to the temperature above the melting point.

After that, relatively rapid cooling is realized, due to which the cell material passes into the amorphous state (RESET).

Currently, alloys based on tellurium, for example $Ge_2Sb_2Te_5$ or Sb_2Te, doped with Ag/In ($Ag_5In_5Sb_{60}Te_{30}$) are used as the working layer of PCM memory elements [1]. However, these chalcogenide alloys have a relatively low crystallization temperature; therefore problems may occur in connection with thermal stability and data integrity in the amorphous state. In addition, tellurium has a high vapor pressure and a low melting point, which complicates the technological process of manufacturing devices. There are undesirable interactions of materials leading to inhomogeneity in the working layer composition and reduction of the cells service life. Therefore, the properties of non-chalcogenide alloys have been actively studied, for example Ge-Sb [2], Ga-Sb [3], Al-Sb [4].

This paper investigates the processes of the structure formation of copper-gold nanoalloys with the goal to analyze their potential of using them as a working layer in PCM memory devices.

2. Computer model

The operation principle of PCM memory devices determines the requirements for the working layer material of cells. Firstly, the resistance of the active elements should depend on the type of the substance substructure. Secondly, a clear distinction is needed between the parameters of the system at which the transition to the amorphous and crystalline phases occurs. Thirdly, these states should be stable, i.e. there should be no spontaneous phase transition during the process of reading the information and its long-term storage. Fourthly, trends in the improvement of nonvolatile memory devices mean an increase in the density of data recording. Therefore, small nanoclusters with diameters up to 10 nm are of interest for study.

In this work nanoclusters of Cu-Au alloy with diameters of 2.0, 4.0, 6.0, and 8.0 nm were studied with computer simulation. For this purpose the molecular dynamics method was used, which describes quite well a wide range of properties of various metals and alloys. The interatomic interaction forces were calculated with the modified tight-binding potential (TB-SMA). The average kinetic energy of atoms was defined with the rapid Verlet algorithm (time step h = 1.0 fs). Then, based on this, the system temperature was calculated.

The simulation was carried out with the MDNTP computer program developed by Dr. Ralf Meyer (University Duisburg, Germany).

The initial spherical clusters were obtained by cutting out them from the copper lattice with the ideal FCC structure. Then a part of Cu atoms in them was randomly replaced with Au atoms in certain percentage ratios. After that all the clusters underwent the compulsory procedure of thermal relaxation at the temperature of 100 K.

3. Results and discussion

The chemical composition of binary nanoclusters as well as their size affects significantly their properties. In some cases even the addition of a single impurity atom can greatly change the particle characteristics. In case we consider these clusters from the standpoint of using them as a working layer in PCM memory devices, the cooling time of the particles of the melt is added to the first two factors since the processes of formation of amorphous and crystalline configurations are associated with a more intense or lower rate of heat removal respectively.

In this paper we investigated all the above mentioned factors. The cluster diameters varied from 2.0 to 8.0 nm. The chemical composition of the clusters was changed, namely the proportion of gold in the alloy varied from 90 to 10, 70 to 30, 50 to 50, 30 to 70, and from 10 to 90. At the first stage the simulation was carried out with the Nosé thermostat. The clusters were stepwise heated to the temperature of 1500 K (with a step of 50 K) until the complete destruction of the long-range order in them, and at each fixed value they were kept for 0.3 ns. At the second stage of the experiment the studied ensembles were smoothly cooled with the Andersen thermostat to the temperature of 300 K, with some fixed speeds corresponding to the cooling periods of 0.4, 1.0, and 4.0 ns (or the heat removal rate of $30 \cdot 10^{11}$, $12 \cdot 10^{11}$, $3 \cdot 10^{11}$ K s^{-1}). The simulation results are presented in table 1.

Table 1. Probabilities of the formation of amorphous and crystalline structures of Cu-Au nanoclusters of different chemical compositions with varying cluster diameters d and the heat removal rate $\Delta T\ \Delta t^{-1}$.

d (nm)	$\left(\Delta T\ \Delta t^{-1}\right)\cdot 10^{11}$ (K s^{-1})	Probabilities of crystalline structure formation (%)				
		$Cu_{90}Au_{10}$	$Cu_{70}Au_{30}$	$Cu_{50}Au_{50}$	$Cu_{30}Au_{70}$	$Cu_{10}Au_{90}$
2	30	70	40	0	0	10
	12	100	40	0	0	0
	3	100	90	30	40	100
4	30	30	0	0	0	0
	12	60	0	0	0	0
	3	100	80	30	20	100
6	30	0	0	0	0	0
	12	60	0	0	0	0
	3	90	70	20	0	100
8	30	0	0	0	0	0
	12	50	0	0	0	0
	3	100	60	0	0	100
d (nm)	$\left(\Delta T\ \Delta t^{-1}\right)\cdot 10^{11}$ (K s^{-1})	Probabilities of amorphous structure formation (%)				
		$Cu_{90}Au_{10}$	$Cu_{70}Au_{30}$	$Cu_{50}Au_{50}$	$Cu_{30}Au_{70}$	$Cu_{10}Au_{90}$
2	30	30	60	100	100	90
	12	0	60	100	100	100
	3	0	10	70	60	0
4	30	70	100	100	100	100
	12	40	100	100	100	100
	3	0	20	70	80	0
6	30	100	100	100	100	100
	12	40	100	100	100	100
	3	10	30	80	100	0
8	30	100	100	100	100	100
	12	50	100	100	100	100
	3	0	40	100	100	0

The experiment with pure substances has shown that the process of the structure formation of Cu and Au is very much dependent on the cooling conditions [5]. The crystalline modifications predominate at a low rate of heat removal. The amorphous structures prevail with intensive cooling. Moreover, the increase of the cluster diameter stabilizes the formation of the crystal structure of both copper and gold. Here arise the problems of obtaining the probability of amorphous modification necessary for the functioning of PCM materials. To achieve these parameters by increasing the cooling rate is not rational because of a number of serious difficulties in their technical implementation. Therefore, the purpose of simulating Cu-Au nanoclusters was to search for the configurations where the appearance of amorphous structures becomes stable with the increase in the heat removal rate, while at the same time there are sound indicators of the formation of crystalline modifications during slow cooling.

The analysis of the results shows that the amorphous configuration is preferable for $Cu_{50}Au_{50}$ and $Cu_{30}Au_{70}$ nanoclusters, regardless of the heat removal rate. Moreover, the increase in the particle diameter makes this dependence more pronounced. Alloy $Cu_{70}Au_{30}$ generally shows similar results. Thus, at the heat removal rate of $\Delta T\ \Delta t^{-1} = 3\cdot 10^{11}$ K s^{-1} the proportion of the crystalline structures decreases from 90 % to 60 % when the cluster size is growing. $Cu_{90}Au_{10}$ particles show a wider range of values. Here, at $\Delta T\ \Delta t^{-1} = 30\cdot 10^{11}$ K s^{-1} the amorphous configuration prevails (it is stabilized at

d = 6.0 nm), while at $\Delta T\,\Delta t^{-1} = 3\cdot 10^{11}$ K s^{-1} the crystalline one predominates (the maximum is reached at d = 8.0 nm). However, the most interesting is the configuration of $Cu_{10}Au_{90}$. In this case, the particle substructure type doesn't depend on the cluster diameter being determined by the heat removal rate. The crystalline structures are formed during slow cooling (100 % of all the cases), the amorphous compositions being shaped at high and average cooling rates (90 – 100 %). Figure 1 shows typical structural configurations of $Cu_{10}Au_{90}$ nanocluster with the diameter of 2.0 nm.

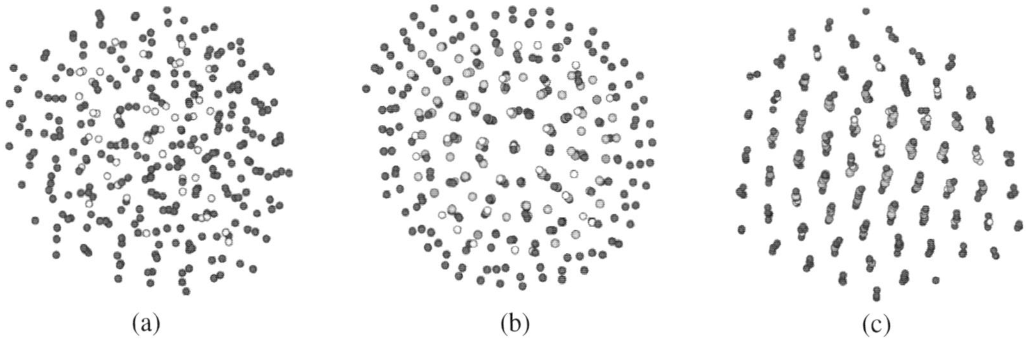

Figure 1. The typical structural configurations of $Cu_{10}Au_{90}$ nanocluster with the diameter of 2.0 nm: (a) amorphous ($\Delta T\,\Delta t^{-1} = 30\cdot 10^{11}$ K s^{-1}); (b) icosahedral ($\Delta T\,\Delta t^{-1} = 3\cdot 10^{11}$ K s^{-1}); (c) decahedral ($\Delta T\,\Delta t^{-1} = 3\cdot 10^{11}$ K s^{-1}).

Thus, according to the research results, it can be concluded that $Cu_{90}Au_{10}$ and $Cu_{10}Au_{90}$ alloys can act as the PCM memory cell material when the cluster size is at least 8.0 nm in the first case and 4.0 nm in the second one.

4. Conclusion

The improvement of phase charge memory devices is associated with researches of the properties of different tellurium-based alloys and non-chalcogenide compounds of germanium, hafnium and plumbum as well as with the investigation of new materials for the active layer of cells. The metal nanoclusters are of particular interest for study. This is primarily due to their relatively low cost and high probability of obtaining particles with the required size distribution during the synthesis process.

This work investigated the potential of using Cu-Au nanoclusters as a working layer of PCM memory cells. The smooth cooling of the melt of Cu-Au nanoclusters of 2.0 – 8.0 nm diameter was simulated with the molecular dynamics' method, using the modified tight-binding potential (TB-SMA). As a result of this, we identified two nanoalloys – $Cu_{90}Au_{10}$ and $Cu_{10}Au_{90}$. The first alloy has demonstrated a significant influence on the size effects. The formation of the amorphous structure is stabilized with the increase in the cluster diameter. Thus, the application of $Cu_{90}Au_{10}$ and $Cu_{10}Au_{90}$ nanoalloys as bits in PCM memory devices is possible; though there are limitations on the minimum cluster size of 8.0 and 4.0 nm, respectively.

To understand the nature of the structural transitions of copper-gold alloy nanoclusters, it is necessary to study a wider range of cooling rates and conduct a more detailed structural analysis of the particles.

Acknowledgments

The presented work was carried out with the financial support of the Russian Foundation for Basic Research, grant numbers 18-42-190001 and 19-48-190002.

IOP Conf. Series: Materials Science and Engineering **684** (2019) 012014 doi:10.1088/1757-899X/684/1/012014

References

[1] Noel P, Vallee C, Hippert F, Fillot F and Raty J-Y 2017 *Semicond. Sci. Technology* **33** 1–54
[2] Krusin-Elbaum L, Shakhvorostov D, Cabral Jr. C, Raoux S and Jordan-Sweet J L 2010 *Appl. Phys. Lett.* **96** 121906
[3] Velea A, Borca C N, Socol G, Galca A C, Grolimund D, Popescu M and van Bokhoven J A 2014 *J. of Appl. Phys.* **116** 234306
[4] Zhou X, Wu L, Song Z, Rao F and Ren K 2013 *Appl. Phys. Lett.* **103** 072114
[5] Gafner Yu Ya, Gafner S L, Redel L V and Golovenko Zh V 2014 *J. Nanosci. Nanotechnol.* **14** 5138–44

IOP Conf. Series: Materials Science and Engineering **684** (2019) 012015 doi:10.1088/1757-899X/684/1/012015

3D aerosol printing as method to obtain low-temperature ceramic layers of printed circuit boards

A A Ivanov[1] and A S Chermoshentseva[2]

[1] Tomsk State University of Control Systems and Radioelectronics, 40 Lenina Avenue, Tomsk, 634050, Russia
[2] Bauman Moscow State Technical University, 5 building 1 2nd Baumanskaya St., Moscow, 105005, Russia

e-mail: alexchemtsu@rambler.ru

Abstract. A method for the synthesis of the polyaluminosilicates (PAS) with the macromolecules of dendrimer morphology has been developed. A quantum-chemical assessment of thermodynamically possible mechanisms to develop the populations of short-chain aluminosilicates (nuclei of dendrimer macromolecules) has been carried out followed by the domains generation from the dendrimer assemblies capable to contain up to 80 % wt of the dispersed filler particles into their interdomain spaces by means of the GAUSSIAN B3LYP/6-31G (d) software package. A quantitative calculation of the parameters determining the composition of PAS macromolecules and the size of the dendrimer assemblies has been carried out. A method for the synthesis of monomeric precursor-powder has been developed. A new low-temperature ceramic composite material for 3D aerosol printing has been developed on the basis of the PAS and highly dispersed precursor-powder. The processes of 3D aerosol printing of the PAS filled with monomeric precursor-powder have been researched. An optimal method for aerosol generation has been chosen that does not affect the PAS supramolecular structure. The ceramic dielectric layers with high consumer properties have been obtained by means of 3D aerosol printing.

1. Introduction

Microelectronic products and spacecraft devices with a long active service period based on multifunctional structural materials can be successfully developed provided that the new generation polymer composite material is created. The main problem is the heat dissipation while using such products. The importance to develop materials with high thermal conductivity, resistance to aggressive environments, high temperatures and their significant differences, various radiations, high vacuum, and high pressures becomes more urgent with increasing human activity.

Nowadays there is an important task to apply such new composite materials on the surfaces of various products and devices along with the problem to develop them in many industries. 3D printing is the modern method to form various coatings and topologies in different products and devices, including microelectronics.

Nowadays it is not always possible to obtain the homogeneous nanostructured ceramic coatings with a specific microstructure, high purity, high thermal conductivity, and simultaneously with a rational stoichiometry. That is why the development of microprinting technology is limited to the technologies for producing consumable products with the necessary functional and structural properties, as well as a thorough physicochemical and morphological analysis of the obtained polymer matrices and their components.

The role of the controlled chemical synthesis of the substances for ceramic materials that form multilayer coatings is decisive to solve the problem. The reason is that nanosize, phase, structural, and other features that determine the physical, mechanical, chemical, and operational properties of end coatings for microelectronic products and spacecraft devices with a long active service period under severe operating conditions form at the synthesis stage. This problem is to a large extent limited to the synthesis and chemical modification of various functionalized nanoparticles with the set chemical and phase composition and morphology and also to the study of such particles behavior in the initial matrix and in the process of solvent removing from a limited volume of a microdroplet [1].

Nowadays many high-tech companies all over the world switch to the technology of prototyping and commercial production of modern microelectronic and spacecraft devices using computer design and microprinting. It is not possible to obtain a specific dielectric layer or topology with the set thermophysical, physical, and chemical properties at the certain area of a device or product by means of traditional methods with the use of commonly applied materials because metallic and nonmetallic materials commonly applied in various industries have to a great extent reached their functional and structural limits.

Physical chemistry of high-molecular compounds is rapidly developing in the frame of fundamental science and different technologies. It is a new field concerned with the synthesis and study of the structure and properties of three-dimensional hyperbranched polymers and oligomers, the so-called dendrimers (fractals) [2]. The materials with such macromolecular assembly morphology attract our attention because the number of branches increases exponentially by each chain growth act in three-dimensional macromolecular assembly synthesis. As a result, the macromolecule size and shape change that leads to major physical and physicochemical properties change such as viscosity, solubility, moisture absorption, density etc.

Thus, special synthetic approaches allow obtaining regular dendrimer assemblies (DA), the macromolecules of which have a well-defined molecular weight. Furthermore, it should be noted that a lot of physical and physicochemical properties of dendrimer materials such as glass transition temperature mainly depend on the chemical nature of the terminal groups located on such spherical macromolecules-fractals.

The modern and available method to produce functional and structural materials with the necessary properties is currently a sol-gel technology that allows controlling the chemical nature of the material in the synthesis phase, thereby setting the required properties for the finished composites. The sol-gel technology is the primary method to obtain multipurpose nanodispersed materials. The actual tendency of sol-gel synthesis is to modify the siliceous compounds by both organic and inorganic substances. The sol-gel method peculiarity is the hydrolytic polycondensation reaction of alkaline silicates that occurs in the presence of inorganic substances (salt, acid) and low and high molecular organic modifiers. They give the specified physicochemical and technically essential properties to the formed materials. As a rule, the inorganic substances and modifiers are template agents contributing to the formation of silicate and hybrid nanocomposites with a specific structure [3-8].

The primary task is to apply the obtained material on different in their chemical nature bases and study the processes that occur when applying the obtained material. This task is to be solved along with the basic task to produce new ceramic composite materials with specified functional and structural properties used as consumable products for 3D printing of various coatings and topologies in microelectronic devices and products. Such process task as layers and topologies application generates current scientific challenges of surface physical chemistry, for example, the problem of material adhesive power and strength. Simultaneously, applying material on the product or device surface generates fundamental research of physicochemical processes.

Currently there are many methods of additive printing of composite materials that relate to printing technology. The printing technology development is based on the achievements in the field of nanotechnology and nanomaterials. Consequently, the production of functional elements with the use of a printer relates to nanoelectronics. Research and development in this field of nanotechnology is a crucial task because it is dynamically developing now.

2. Experiment

2.1. Composite material obtaining

One of the study objectives related to physical chemistry of high-molecular compounds is to study how the PAS chemical structure influences on the macromolecule sizes and conformations. The molecular dendrimer (fractal) system is a regularly branched tree-shaped structure. The main unit of their macromolecules are crowns (dendrons). The combination of crowns in one central unit leads to the formation of dendrimers, the macromolecules of which have a close to spherical shape (figure 1). The crowns inclusion in the repeating units of linear polymers allows obtaining fractal polymers or, as they are often called, cylindrical dendrimers. The possibility to obtain the terminal groups of macromolecules controlled by chemical structure, size, and number, as well as to modify them by varying the crown generation number and physicochemical structure of the terminal fragments allows solving the fundamental problems of physical chemistry of complex branched molecular systems and developing new functional materials for microelectronics. It seems to be important to research comprehensively the influence of the generation number, chemical structure of the crown and end groups on the sizes and conformations of dendrimer macromolecules and crown-modified polymers in the polymer systems filled with highly dispersed fillers.

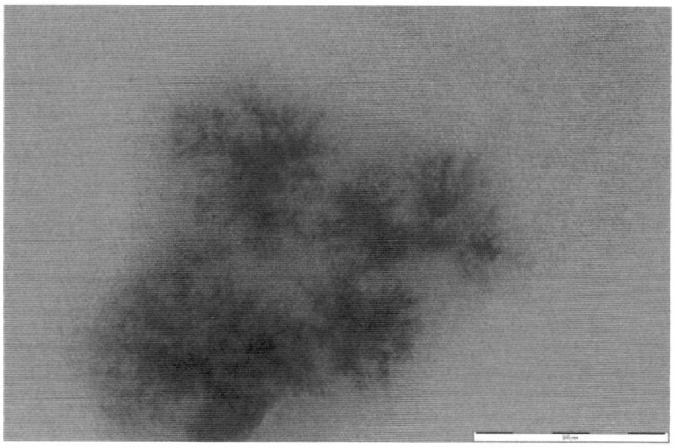

Figure 1. TEM photograph of dendrimers of PAS macromolecular assemblies (200 nm scale, Philips CM 30 transmission electron microscope).

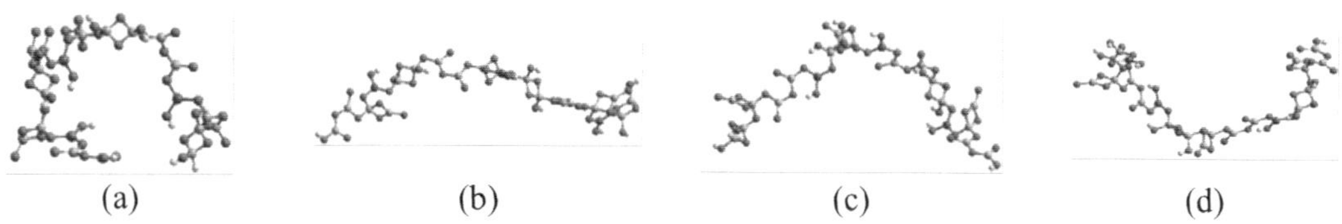

Figure 2. Models of oligomer PAS conformers that are able to play role of nuclei in dendrimer assemblies and participate in crown formation (visualized by GAUSSIAN computer program). Viscosity has been measured by Brookfield DV2TRV viscometer. (a) – viscosity 1560 cPs, (b) – viscosity 2134 cPs, (c) – viscosity 1873 cPs, (d) – viscosity 1947 cPs.

A silicic acid and aluminum nitrate have been chosen as starting reagents to form the dendrimer PAS assemblies for the dendrimer PAS synthesis to obtain the filled tailored materials. The formation has been carried out in three stages: I stage: the silicic acid dissolution by the pH > 7; II stage: the aluminum nitrate hydrolysis; III stage: the light polycondensation with gel-precursor formation from oligosilicic acids and $O_3N\text{-}Al\text{-}(OH)_2$ [3].

Depending on the number of initial components used for the synthesis and polycondensation reaction temperature we have obtained the PAS with a different structure of the polymer chain (figure 2) having different viscosity (characterized by a molecular weight) and able to accommodate different amounts of dispersed filler in their interdomain spaces.

A simple comparison of the visualized models images (calculated by the quantum-chemical method) (figure 2) and the microphotograph of the dendrimer PAS assembly obtained by a transmission electron microscope (TEM) (figure 1) allows us to conclude that the morphological picture of the obtained PAS is identical with the one obtained by TEM and the participation of oligomer segments in crown formation. In other words, the formation of PAS macromolecules begins with the formation of a dendrimer nucleus, on which other arcuate oligomer segments are grown in spatially accessible OH-groups. As a result, the macromolecules form, i.e. the dendrimers that are clearly seen on the TEM image (figure 1).

The study of the PAS chemical structure begins with the study of viscosity characteristics and the effect of the dendrimer macromolecule generation number on the behavior of the terminal groups that form the crown of a PAS domain. The limiting sizes, configuration, and molecular weight characteristics of regular dendrimers can be theoretically predicted by means of the quantum-chemical assessment of the conformations of the initial PAS oligomer compounds (figure 2).

Since the growth of crown generations in all directions is equally probable, they take a close to spherical shape after 3 or 4 generations. However, it should be taken into account that the crown thickness increases with an increase in the generation number of the dendrimer macromolecule. The crown size increasing is characterized by the unfolding of the terminal chains in the radial direction of the macromolecules. Such a steric effect is possible due to the interactions occurring between the terminal chains due to the dendritic migration surface area decrease per one crown as they increase in the macromolecule. Consequently, an increase in the generation number G stimulates the terminal groups self-organization that tend to transit to the PAS dendrimer macromolecule outer layer. Then since all the macromolecules increase in their molecular weight by the same amount at each stage of the PAS controlled synthesis, an almost monodispersed PAS forms as a result.

2.2. The synthesis of monomeric precursor-powder

The synthesis of monomeric precursor-powder has been carried out by doping. It is doping that is one of the ways to obtain powders with a nanosized structure at the lowest cost. The synthesis is based on the chemical plating method.

At the first stage, $Al(NO_3)_3 \times 9H_2O$ has been chosen as an initial reagent. The method of reverse plating has been used to ensure the maximum completeness of plating components and high degree of the plate homogeneity.

The following points have been taken into account by the plating:

- the diluted solutions of salts and a precipitator have been used (~0.12 M);
- the feed rate of the salt solutions has been chosen so that pH during the plating is ~9.2;
- plating has been carried out at a temperature of minus 6 °C and with thorough stirring;
- the plating time in the mother solution has been minimal after the plating completion in order to slow the growth of agglomerates.

The second stage of the synthesis has been the ultrasonic treatment of the obtained plates in order to decrease their agglomeration degree. Then the deagglomerated plates have been processed at a temperature of minus 25 °C (20 hours) that allowed carrying out the deep dehydration of the plates and preserve the formed nano-dimensions of the co-plated powders. According to the sedimentation analysis, the agglomerate size in the co-plated powders is 20–40 nm after freezing (figure 3).

Figure 3. Microphotograph of nanosized Al_2O_3 precursor-powder microstructure after deagglomeration (magnification is 95000).

2.3. The monomeric precursor-powder injection into the PAS matrix

After viscometric measurements of the synthesized PAS, the distances between the chain ends of the macromolecules segments L have been calculated. The conformer geometry and shape have been assessed by the quantum-chemical optimization methods of the geometry and energy of possible molecular assemblies-segments that are the products of PAS oligomerization. The distances $L = 15.11$ Å (angstroms) for the conformer *a* (figure 2(a)), the distances $L = 37.13$ Å for the conformer *b* (figure 2(b)), the distances $L = 28.39$ Å for the conformer *c* (figure 2(c)), distances $L = 28.66$ Å for the conformer *d* (figure 2(d)). The conformer *a* is the most probable conformer involved in strong branching (crown formation) of the aluminosilicates, since the crown size increase is characterized by the unfolding of the end chains in the radial direction of the macromolecules. Although the distance between the ends of the conformers *c* and *d* is only 1.8 times larger than that of the conformer *a*, the process of PAS polycondensation with the formation of linear macromolecules is difficult. Consequently, the results of the quantum-chemical calculations correlate with the data of the experimentally obtained viscometric measurements.

One controls the supramolecular dendrimer assemblies by regulating the quantitative characteristics defining their microstructure and size according to certain parameters, such as the number of spatial segments of branches from generations or crowns (N_c), the number of branches formed by each spatial segment or the branching index of links (N_b), the number of terminal groups (Z), the number of generations (G).

We have analyzed the dendrimer supramolecular formations of the synthesized aluminosilicates according to the scheme (figure 4) and have calculated the polymerization degree (P), N_c, N_b and G for the conformer *a* as for the most probable one. The values obtained for the conformer *a* are the following: $N_c = 3$, $N_b = 3$ and $G = 3$. The polymerization degree P or number of repeating units with $N_c = 3$, $N_b = 3$ and $G = 3$ have been calculated according to the equation (1) that has the value 120 (P = 120).

$$P = N_c \cdot \frac{N_b^{G+1} - 1}{N_b - 1} \tag{1}$$

The number of terminal groups is 81 ($Z = 81$) and has been calculated according to the equation (2).

$$Z = N_c \cdot N_b^G \tag{2}$$

The molecular weight of the amorphous dendrimer PAS M is ~480000; $M_t = M_u$ ~ 1300. The value M has been calculated according to the equation (3).

$$M = M_c + N_c \cdot \left[\frac{M_u \cdot \left(N_b^{G+1} - 1 \right)}{N_b - 1} + M_t \cdot N_b^{G+1} \right], \quad (3)$$

where M_c – the molecular weights of the nucleus, M_u – repeating unit formed from the initial compounds and M_t – terminal group.

The regularity violations in the structure of the supramolecular PAS formations and the lack of predictability may be caused by the traditional reaction mechanism violation, such as the presence of side intramolecular reactions, incomplete targeted reactions due to steric difficulties arising from the substituents accumulation etc. However, it is considered that the branching regularity violations in dendrimer macromolecules do not dramatically affect their size. This is confirmed by the experiments, in which the dendrimers structure regularity has been deliberately violated. The same conclusion can be made from the data on the measurement of the macromolecules size of the dendrite-like polymers obtained by the method of one-stage synthesis.

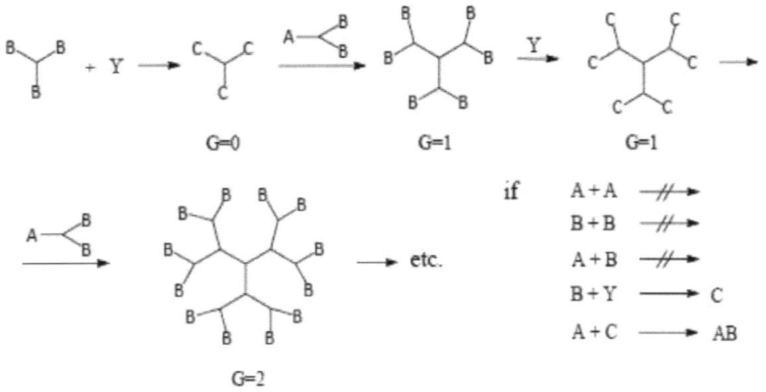

Figure 4. Scheme of controlled synthesis of dendrimers.

We have analyzed the elemental composition of the dendrimer PAS by the X-ray fluorescence spectroscopy to confirm quantitatively the calculated degrees of the polymerization and molecular weight (table 1).

Table 1. Elemental composition of the dendrimer PAS (%).

Element	X-ray fluorescence analysis (%)	Computer experiment (%)
O	65.13	67.67
Si	25.67	27.68
Al	7.07	4.61
K	0.49	
Na	0.47	
Ca	0.23	
Zr	0.21	

The content of O and Si elements in the model visualized samples of dendrimer PAS is close to the experimental values. The difference in Al content differs by a factor of 1.5. Perhaps, it can be explained by the discrepancy in the weights of this element in the experimental and theoretical samples. Thus, the choice of oligomer conformer segments to calculate the sizes of dendrimer PAS should be considered reasonable.

One interrupts stage III to obtain the filled AS samples so that the well-proportioned and adequate dendrimer crowns form (see the stage-to-stage scheme of the PAS synthesis). The monomeric precursor-powder nanoparticles are injected in the crowns by mechanochemical mixing with simultaneous ultrasonic treatment. The highly dispersed filler powder has been injected by means of the Emax high-speed ball mill of Retsch company and followed by ultrasonic dispersing. The ultrasonic dispersing of particles is the inclusion of monomeric precursor-powders exhibiting weak van der Waals multiparticle interactions. The monomeric precursor-powders are injected into the voids of the PAS dendritic structures. The molecules of the monomeric precursor-powder penetrate into the dendritic supramolecular structure and increase the interlayer distance. Then the interaction between the layers weakens, and it becomes possible to disperse the nanosized particles under the action of thermal oscillations.

Thus, we have managed to obtain a composite nanostructured material consisting of the dendrimer PAS and monomeric precursor-powder uniformly distributed in it (figure 5).

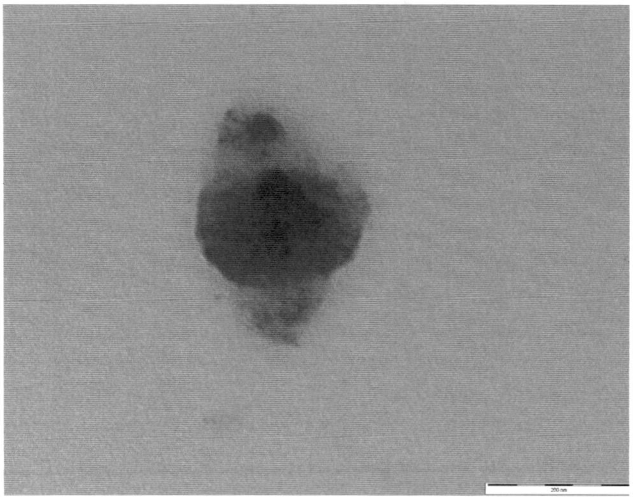

Figure 5. Microphotograph of PAS with monomeric precursor-powder injected in it (200 nm scale, Philips CM 30 transmission electron microscope).

Figure 5 shows that the PAS dendritic supramolecular structure is densely filled with the monomeric precursor-powder.

The use of the dispersed filler powders obtained by other methods with a layered (scale) structure to inject into the PAS matrix is difficult due to the possibility lack to inject them into the dendrimeric PAS interdomain spaces.

3. Results and discussion

3.1. Ceramic layer printing on an aluminium base by means of 3D aerosol printing

One of the most crucial problems solved in some technological processes and definitely in 3D aerosol printing is to obtain aerosol from dispersed liquids.

The dispersing can be carried out by various methods, the most widespread of which are hydraulic, mechanical, pneumatic, and electrostatic. We have considered the advantages and disadvantages of each method separately in our previous studies.

According to the research results, the most preferable method for us is the pneumatic one that allows working with highly viscous materials without destroying the supramolecular structure of a polymer binder.

IOP Conf. Series: Materials Science and Engineering **684** (2019) 012015 doi:10.1088/1757-899X/684/1/012015

We have used the synthesized polyaluminosilicate filled with a highly dispersed filler and the Aerosol Jet 15EX 3D printer of Neotech AMT Company to obtain ceramic layers (figure 6) and the topologies on aluminum bases by 3D aerosol printing with the help of the pneumatic method.

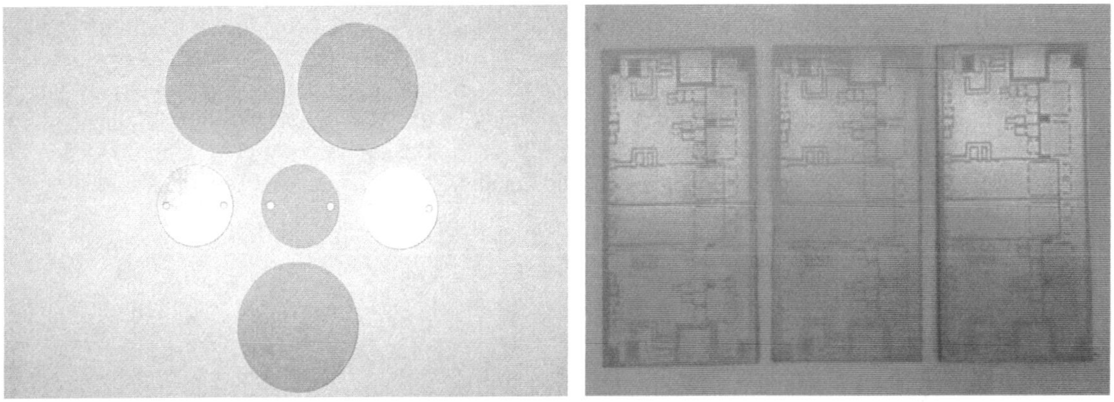

Figure 6. Photographs of metal base samples with dielectric layers applied by 3D aerosol printing.

3.2. Study of the ceramic layers durability on an aluminum base obtained by 3D aerosol printing
A track has been formed on the ceramic coating by means of a high-temperature tribometer with the following specified measurement parameters:

- the track radius is 3.02 mm;
- the applied load is 10 N;
- speed is 5 cm s^{-1};
- the total path traveled is 114 meters.

The durability dependence (the penetration depth of the instrument holder head) on the traveled path has been obtained according to the tribological reserach results (figure 7). An optical profilometry of the track has been carried out after the tribological research.

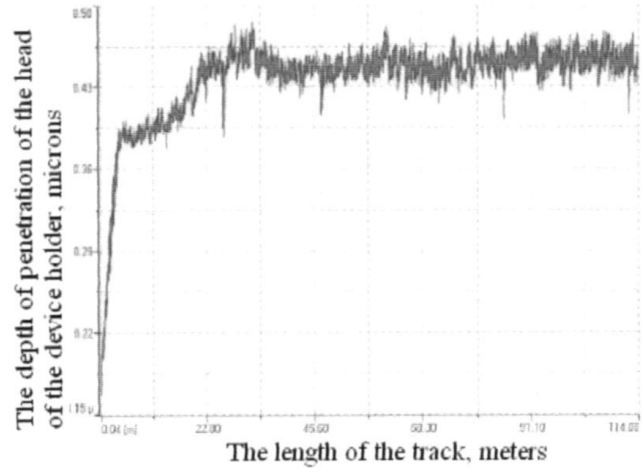

Figure 7. Dependence of penetration depth of instrument holder head on traveled path length by sample with ceramic coating on aluminum base.

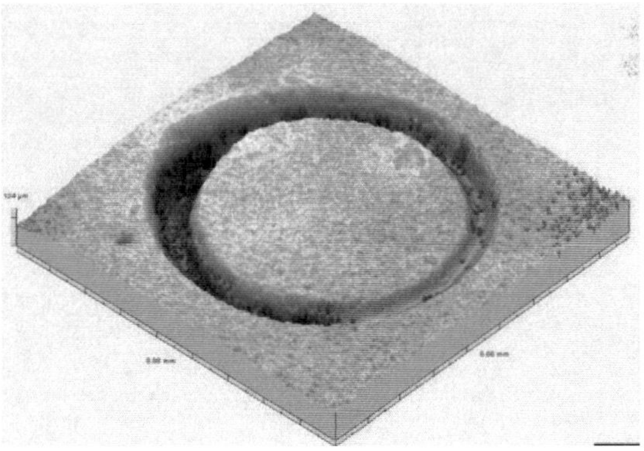

Figure 8. Track surface 3D image of sample with ceramic coating on aluminum base after tribometric research.

Figures 7 and 8 show that the coating abrasion along the track occurs evenly and without delamination that has a positive effect on the surface treatment of the coatings formed from filled dendrimer PAS in end products and microelectronic devices.

4. Conclusion

The polyaluminosilicate of dendrimer morphology has been synthesized during the research. The most probable conformers to form the dendrimer PAS have been determined by the quantum-chemical assessment of the thermodynamically possible mechanisms to form the populations of short-chain aluminosilicates. The viscometric measurements allowed us to confirm the quantum-chemical calculations and modify the mechanism to obtain the dendrimer PAS.

The synthesis method for the monomeric precursor-powder, the structure and morphology of which allowed obtaining a PAS close-packed polymer supramolecular structure, has been developed to obtain the filled polyaluminosilicates used as materials for 3D aerosol printing. The ceramic dielectric layers with high consumer properties have been obtained with the use of the obtained materials by means of 3D aerosol printing.

Acknowledgments

The reported study was funded by RFBR according to the research project no. 18-29-11018\18.

References

[1] Ivanov A A, Polyshko V A and Khomyakov A Yu 2019 *IOP Conf. Ser.: Mater. Sci. Eng.* **510**
[2] Fritz V, Richardt G and Werner N 2009 *Dendrimer Chemistry: Concepts, Syntheses, Properties, Applications* (Weinheim: WILEY-VCH Verlag GmbH & Co. KGaA)
[3] Ivanov A A, Botvin V V and Filimoshkin A G 2014 *Russ. J. Appl. Chem* **87(2)** 141–50
[4] Park J U et al. 2007 *Nature Materials* **6(10)** 782–9
[5] Takahashi K et al. 2006 *Int. J. of Adhesion and Adhesives* **26** 639–43
[6] Tomalia D A and Durst H D 1993 *Top. Curr. Chem.* **165** 193–313
[7] Tomalia D A, Naylor A M and Goddard W A 1990 *Angew Chem.* **102** 119–57
[8] Tomalia D A et al. 1986 *Macromolecules* **19** 2466–8

IOP Conf. Series: Materials Science and Engineering **684** (2019) 012016 doi:10.1088/1757-899X/684/1/012016

On improving the operational properties of metal-fluoroplastic plain bearings

V E Rogov and A B Baldanov

Baikal Institute of Nature Management, Siberian Branch of Russian Academy of Sciences, 6 Sakhyanovoy Street, Ulan-Ude, 670047, Russia

e-mail: rogov54v@mail.ru

Abstract. The analysis of metal-fluoroplastic materials was carried out. Domestic metal-fluoroplastic materials do not meet the requirements of regulatory and technical documentation adopted in domestic engineering. A review of patents on technologies for the production of metal fluoroplastic materials (MFM) was carried out. Promising methods for producing MFM have been identified. New methods can significantly increase the operational characteristics of fluoroplastic materials. For widespread use of metal-fluoroplastic plain bearings it is necessary to increase their performance. Reducing the linear wear parameter is possible by increasing the nominal area of the metal contact.

1. Introduction

To run without a lubricant, plain bearings made of metal-fluoroplastic materials are used in modern machines. Currently, metal-fluoroplastic bearings are understood as a fairly wide range of products made of metal with polytetrafluoroethylene (PTFE). These include porous metal frames impregnated with fluoroplastic and various tape materials on a steel basis with a porous antifriction surface layer. A porous surface is necessary to hold the fluoroplastic on the surface and in the pores, since it has low adhesive properties to any surfaces. The production of bearings from the MFM was established in the last century [1]. They are used in friction units of modern technical devices – cars, airplanes and marine vessels [2]. The appropriateness of their use is due to their lightness, variability in composition, structure and a wide range of technical properties. Domestic fluoroplastic materials do not fully meet the requirements of regulatory and technical documentation adopted in the domestic aircraft industry [2,3]. Imported IMF products of the DU and DP brands are widely used in various industries and aircraft manufacturing. The lack of these materials with high performance characteristics not inferior to imported counterparts adversely affects the development and production of the northern performance equipment under the state program "Socio-economic development of the Arctic zone of the Russian Federation." In this regard, it is necessary to identify promising technologies that allow establishing the production of these materials with high performance not inferior to foreign analogues.

2. Literature review

The traditional MFM module, its production method and characteristics have been detailed in [4]. MFMs have high antifriction properties in the temperature range from 73 to 553 K; pV = 1.5 MPa-m s^{-1}. The ultimate wear in such a bearing is concentrated in the surface layer with a thickness of

0.025-0.05 mm. This wear parameter affects the durability of the product as a whole [5]. Low reliability is explained by the insufficient amount of fluoroplastic located in the porous space of bronze particles [6]. The material is used in units operating in the modes of hydrodynamic and mixed friction (in the friction units of pumps for pumping oil, various rotary mechanisms, hydraulic boosters etc.). Metal-fluoroplastic materials and products are produced in the Russian Federation at machine-building enterprises such as Kompozit LLC, Promsnabkomplekt LLC (Kineshma), Ftoroplast LLC (Bugulma).

The enhanced performance of foreign-made MFM plain bearings is conditioned by technology. Abroad, external pressure is used during the heat treatment of the polymer layer in the porous layer [6]. This technology allows the polymer to be fixed in the pores, which cannot be achieved with free sintering in air. It is known that a necessary condition for the performance of plain bearings is the optimal combination of the amount of polymer and metal in the structures [7].

Thus, in order to create modern plain bearings operating without a lubricant, it is necessary to increase the thickness of the polymer working layer, fixing the fluoroplastic coating-lubricant at this should be carried out under pressure. An increase in the thickness of the polymer layer can be achieved due to the larger pore volume. As the porous layers, various matrix materials can be used. However, traditional matrix-filled materials are expensive to obtain the necessary relief on the metal surface of the product [8]. According to the authors of [4, 6] to increase the amount of polymer in the porous space it is promising to use metal meshes from antifriction materials. Methods are known for producing MPM in the form of plates [9–11], where instead of the traditional porous bronze layer of spherical particles, various mesh woven materials are used. Anti-friction nets are specially fixed on a metal surface. The method of obtaining these materials is presented quite in full by [4], and we will not describe it. We will focus only on the process of fixing PTFE in the pores of the mesh space. The process of fixing or monolithization of the polymer in the space of the meshes is carried out during heat treatment of workpieces in a limited volume. In the process of heating the polymer in an enclosed space, pressure arises due to the thermal expansion of the fluoroplastic, resulting in monolithization of PTFE particles due to cohesive interaction. The resulting materials are a metal composites with a heterogeneous structure. The working surface of the bearing is a regularly repeated sections of a solid base (bronze mesh) and polymer. The mesh material has a relatively low coefficient of friction. During friction the polymer is capable of forming a protecting from setting intermediate layer in the contact zone. The polymer is evenly distributed throughout the thickness of the mesh. The material has high wear-resistant properties that make it operable in a wide range of temperatures and loads under dry friction at $pV > 3$ MPa-m s^{-1} [12]. This MFM has not regrettably gained acceptance in friction units. According to the authors of [12], the main drawback of such bearings is that at the initial moment of operation the counter body comes into contact with a minimal contour area. In fact, contact occurs on points. The main load in such a unit is distributed over fragments of a solid surface. It is logical that when a load is applied, intensive wear of the bearing wire will be observed until the compatibility of the assembly is achieved. In most cases, when stable operation of the friction unit is attained, the dimensions of the bearing go beyond the limit value of the linear wear value. In this regard, the bearing fails. For the stable operation of MFM plain bearings with a mesh layer it is necessary to solve the following problems: determine the limits of change in the nominal contact area; find a means of gaining the required area for a particular unit.

3. Experiments and discussion

For experiments a bronze-brass simple weave mesh No. 16 was chosen (the number of wires in the weft per 1 cm is 16, the diameter is 0.25 mm, the number of wires on the basis per 1 cm is 12, the diameter is 0.27 mm). The pore size of rectangular cells is 0.5 and 0.56 mm, respectively (figure 1). When fully filled with polymer, the porosity of the mesh is 70 %.

In [13] a theoretical calculation was made of the change in the contact area depending on the linear wear for a given mesh. The change in the contour area of one weft is shown in figure 2.

IOP Conf. Series: Materials Science and Engineering **684** (2019) 012016 doi:10.1088/1757-899X/684/1/012016

Figure 1. Photo of a bronze mesh.

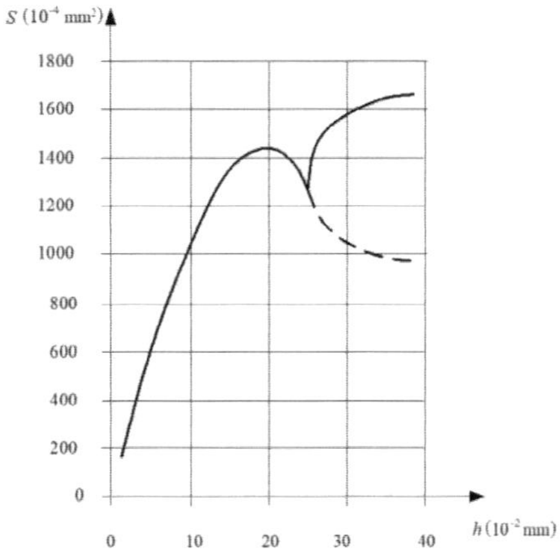

Figure 2. Change in contact area of one weft S depending on the amount of linear wear h.

The graph shows that the contact area of one weft of the mesh varies over a wide interval with a sharp increase of 1000 times or more. Knowing the area of one mesh weft, we can determine the contour area of the bearing. There is an opinion [6] that the contour contact area of meshes can be changed in several ways:

- by the use of meshes with wires of other diameters;
- by the use of meshes with different types of weaving;
- by machining the surface of the mesh.

The calculations and experiments showed that when using meshes of wires of a smaller diameter, their deformation is observed during pressing in the polymer. When using meshes with wires of a larger diameter, the thickness of the bearing and the fluoroplastic consumption increase with a slight increase in area.

When machining the mesh surface of the finished metal-fluoroplastic bearing, an increase in the contour area of each contacting point is observed. However, after machining no polymer antifriction layer is found on the resulting contact points of an ellipse shape. The absence of a dividing fluoroplastic layer on metal contact areas leads to scuffing of metal surfaces when operating without lubricants. Thus, in this case it is necessary not only to increase the contact area, but also to ensure the presence of antifriction polymer on the contacting surfaces. Machining the inner surface of the bearing to size also requires a special tool and its additional fixation.

4. Conclusion

Domestic traditional technologies for the manufacture of MFM do not allow to obtain materials with high performance properties. The reason has to do with an insufficient amount of polymer lubricant located in the pores of bronze. Promising methods are technologies where anti-friction mesh materials are used as a porous layer. The use of nets allows increasing tens of times the thickness of the working polymer layer in comparison with traditional materials. It was revealed that metal-fluoroplastic plain bearings with an antifriction layer in the form of a mesh material do not realize their potential in the absence of a lubricant. The main disadvantage of such materials is the insignificant nominal area of metal mesh fragments in the initial period of operation. Increasing the nominal area in the finished product by its machining removes a significant part of the polymer layer. A possible way to solve this problem is to improve the methods for producing MFM with a mesh woven layer. The required nominal area on the mesh should be obtained by machining during the manufacture of blanks.

Acknowledgments
The work was carried out as part of the state assignment of Baikal Institute of Nature Management, Siberian Branch of Russian Academy of Sciences, and was supported by the state target of the Ministry of Science and Higher Education of the Russian Federation, project No. 9.7667.2017/8.9, and by the grant "Young Scientists of ESSUTM – 2019".

References
[1] Semyonov A P and Savinskii Yu E 1976 *Metal-Fluoroplastic Bearings* (Moscow: Mechanical Engineering)
[2] Bychkov A S, Lavrenko I G, Nechiporenko O Yu, Romashko I M and Mladinov S D 2013 *Open Information and Computer Integrated Technologies* **59** 343–54
[3] Lagunov V S et al. 1996 *Mechanical Engineering* **4** 16–8
[4] Rogov V E and Bokhoeva L A 2018 *Advances in Engineering Research* **158** 343–6
[5] Mashkov Yu K, Ovchar Z N, Baibaratskaya M Yu and Mamaev O A 2004 *Polymeric Composite Materials in Tribotechnics* (Moscow: Subsoil - Business center LLC Press)
[6] Rogov V E, Guriev A M, Nikiforov S O and Kosheleva E A 2010 *Polzunovsky Bulletin* **1** 134–40
[7] 1979 *Metal-Polymer Materials and Products* ed V A Belyi (Moscow: Chemistry)
[8] Belyi V A et al. 1976 *Friction and Wear of Polymer-Based Materials* (Minsk: Science and Technology)
[9] Kornopoltsev N V 1992 *A Method of Manufacturing Multilayer Anti-Friction Products: Russia patent No. 1398244* (Moscow: Russian Federal Service for Intellectual Property)
[10] Kornopoltsev N V 1995 *A Method of Obtaining a Bimetallic Metal Fluoroplastic Material: Russia patent No. 1418999* (Moscow: Russian Federal Service for Intellectual Property)
[11] Kornopoltsev V N, Kornopoltsev N V, Rogov V E, Mognonov D M and Greshilov A D 2002 *A Method of Obtaining a Bimetallic Metal Fluoroplastic Material: Russia patent No. 2212307* (Moscow: Russian Federal Service for Intellectual Property)
[12] Kornopoltsev V N, Kornopoltsev N V and Mognonov D M 2009 *Journal of Friction and Wear* **30(4)** 1–5
[13] Rogov V E and Markhadaev B E 2010 *Assembling in Mechanical Engineering and Instrument-Making* **7** 33–7

IOP Conf. Series: Materials Science and Engineering **684** (2019) 012017 doi:10.1088/1757-899X/684/1/012017

Definition of preferable thickness of metal cylindrical panels with mild camber based on post-buckling behaviour due to compression

O V Mitrofanov

Moscow Aviation Institute (National Research University), 4 Volokolamskoye Highway, Moscow 125993, Russia

e-mail: Oleg1Mitrofanov@yandex.ru

Abstract. Engineer design methodology for metal cylindrical panels with mild camber with post-buckling behavior accepted in case of compression above limit level is suggested. Method is based on geometrically non-linear task analytic solution and is reduced to solution of non-linear equation in panel thickness with limit stress being reached with post-buckling behavior of rigidly supported panel taking into account initial deflection.

1. Introduction

Let us consider load-bearing panel of the aircraft fuselage section with mild camber which is manufactured using welding and permanent strain is possible due to taken design and technology solutions [1]. It should be highlighted that above mentioned permanent strain may be initial deflection (ID) and significantly affect further geometrically non-linear behavior of load-bearing panels in case of loading above limit level. To be definite let us assume that welding along the panel longitudinal axis can lead to ID with transverse waves. According to reference [1] researches results let us assume following parameters of load-bearing panel longitudinal and transverse wave generation: $m = 1$ and $n \geq 1$.

Further on in this investigation let us consider panel minimal thickness definition analytic solution for an early stage of design using geometrically non-linear correlations taking into account initial deflection (ID) [2] for thin isotropic cylindrical panels with mild camber. Let us take into account peculiarities of load-bearing panels design method based on post-buckling behavior suggested in reference [3].

2. Main correlations and statement of the problem

Let us put down main geometrically non-linear correlations of thin isotropic panels. Equation of strain compatibility taking into account initial deflection is of the following form [2]

$$L_1(F) - L_2(W) = 0 \qquad (1)$$

with $L_1(F) = \dfrac{1}{E\delta}\left[\dfrac{\partial^4 F}{\partial x^4} + 2\dfrac{\partial^4 F}{\partial x^2 \partial y^2} + \dfrac{\partial^4 F}{\partial y^4}\right]$,

$$L_2(W)=\left(\frac{\partial^2(W+W_{id})}{\partial x\partial y}\right)^2-\left(\frac{\partial^2 W_{id}}{\partial x\partial y}\right)^2-\frac{\partial^2(W+W_{id})}{\partial x^2}\frac{\partial^2(W+W_{id})}{\partial y^2}+\frac{\partial^2 W_{id}}{\partial x^2}\frac{\partial^2 W_{id}}{\partial y^2}-\frac{1}{R}\frac{\partial^2 W}{\partial x^2}.$$

Let us write non-linear Karman equation of type for isotropic panels taking into account initial deflection of the following form

$$L_3(F,W)-L_4(W)=0 \qquad (2)$$

with $L_4(W)=D\left[\frac{\partial^4 W}{\partial x^4}+2\frac{\partial^4 W}{\partial x^2\partial y^2}+\frac{\partial^4 W}{\partial y^4}\right]$,

$$L_3(F,W)=\frac{\partial^2 F}{\partial y^2}\frac{\partial^2(W+W_{id})}{\partial x^2}+\frac{\partial^2 F}{\partial x^2}\frac{\partial^2(W+W_{id})}{\partial y^2}-2\frac{\partial^2 F}{\partial x\partial y}\frac{\partial^2(W+W_{id})}{\partial x\partial y}+\frac{1}{R}\frac{\partial^2 F}{\partial x^2}$$

with $D(\delta)$ being skin bending stiffness (thickness), F being Airy stress function for which following ratios defining membrane stress are valid:

$$\sigma_x=\frac{1}{\delta}\frac{\partial^2 F}{\partial y^2},\ \sigma_y=\frac{1}{\delta}\frac{\partial^2 F}{\partial x^2},\ \tau_{xy}=-\frac{1}{\delta}\frac{\partial^2 F}{\partial x\partial y}. \qquad (3)$$

Ratios (1) and (2) are initial system of geometrically non-linear equations for post-buckling behavior investigation of isotropic panels with ID. Let us note that there are examples of welding along the skin longitudinal edge with ID formation in reference [1].

Let us consider cylindrical panel with mild camber with ID which can be approximated by following function

$$W_{id}=f_{id}\cdot\sin^2\frac{\pi x}{a}\sin^2\frac{\pi ny}{b}. \qquad (4)$$

Let us present flexure similarly

$$W=f\cdot\sin^2\frac{\pi x}{a}\sin^2\frac{\pi ny}{b}. \qquad (5)$$

After substitution of equations (4)-(5) into non-linear equation of strain compatibility (1) we derive following equation for stress function:

$$F=\left(f^2+2ff_{id}\right)E\delta\left\{A_1\cos\frac{2\pi x}{a}-A_2\cos\frac{4\pi x}{a}+A_3\cos\frac{2\pi ny}{b}-A_4\cos\frac{4\pi ny}{b}-A_5\cos\frac{2\pi x}{a}\cos\frac{2\pi ny}{b}\right.$$
$$\left.+A_6\cos\frac{4\pi x}{a}\cos\frac{2\pi ny}{b}+A_7\cos\frac{2\pi x}{a}\cos\frac{4\pi ny}{b}\right\}+\frac{fE\delta}{R}\left[-A_8\cos\frac{2\pi x}{a}+A_9\cos\frac{2\pi x}{a}\right.$$
$$\left.\times\cos\frac{2\pi ny}{b}\right]+\frac{T_x y^2}{2}$$

with T_x being compression force, $A_1=\frac{n^2a^2}{32b^2}$, $A_2=\frac{n^2a^2}{512b^2}$, $A_3=\frac{b^2}{32n^2a^2}$, $A_4=\frac{b^2}{512n^2a^2}$,

$$A_5 = \frac{n^2 a^2 b^2}{32\left(n^2 a^2 + b^2\right)^2},\ A_6 = \frac{n^2 a^2 b^2}{64\left(n^2 a^2 + 4b^2\right)^2},\ A_7 = \frac{n^2 a^2 b^2}{64\left(4n^2 a^2 + b^2\right)^2},\ A_8 = \frac{a^2}{16\pi^2 R},$$

$$A_9 = \frac{n^2 a^2 b^4}{32\pi^2 R\left(n^2 a^2 + b^2\right)^2}.$$

Using Bubnov-Galerkin method procedure following non-linear equation can be derived:

$$D\frac{\pi^4 f}{4a^3 b^3}\left(3n^4 a^4 + 2n^2 a^2 b^2 + 3b^4\right) + E\delta\frac{\pi^4 n^2}{2ab}\left(f + f_{\text{id}}\right)\left(f^2 + 2ff_{\text{id}}\right)\left(A_1 + A_2 + A_3 + A_4 + A_5 + \frac{A_6}{2}\right.$$

$$\left. + \frac{A_7}{2}\right) + E\delta\frac{\pi^2 \text{b}}{2Ra}\left[\left(A_1 + \frac{A_2}{2}\right)\left(f^2 + 2ff_{id}\right) + \left(A_8 + \frac{A_9}{2}\right)f\right] = \frac{3\pi^2 \text{b} T_{\text{x}}\left(f + f_{id}\right)}{16a}, \quad (6)$$

which associates acting compression force, panel thickness δ, flexure amplitude f and ID. So, with given compression force T_x and known panel thickness flexure amplitude f is defined by equation (6) and membrance stress is calculated using Airy stress function (3). Let us put down equation for longitudinal membrane stress

$$\sigma_{\text{x}} = \frac{1}{\delta}\frac{\partial^2 F}{\partial y^2} = -\frac{4\pi^2 n^2 E}{b^2}\left(f^2 + 2ff_{\text{id}}\right)\Delta_1 - \frac{4\pi^2 n^2 Ef}{Rb^2}A_9\Delta_2 - \frac{T_{\text{x}}}{\delta} \quad (7)$$

with

$$\Delta_1 = \left[A_3\cos\frac{2\pi ny}{b} + 4A_4\cos\frac{4\pi ny}{b} + A_5\cos\frac{2\pi x}{a}\cos\frac{2\pi ny}{b} + A_6\cos\frac{4\pi x}{a}cos\frac{2\pi ny}{b} + 4A_7\cos\frac{2\pi x}{a}\right.$$

$$\left. \times\cos\frac{4\pi ny}{b}\right], \quad (8)$$

$$\Delta_2 = \cos\frac{2\pi x}{a}\cos\frac{2\pi ny}{b}.$$

It should be highlighted that extremum values of functions Δ_1 and Δ_2 do not depend on thickness and load and are defined by correlations of panel geometrical parameters.

3. Engineer design methods

To begin with let us consider design task for thin cylindrical panels with mild camber based on post-buckling state assuming that applied materials design characteristics are known and there is no ID (fid $\approx$ 0). Let us assume that wave generation parameters meet above mentioned conditions that is the most probable panel strain due to taken technology solutions is transverse one. Let us state provisions of engineer method for fuselage section load-bearing skin thickness definition.

At first let us assume that buckling is impossible at limit loading level. In this case thin skins thicknesses $\delta_{\text{stab.}}$ can be defined based on stability conditions.

Secondly with loads above limit level buckling is possible and skin thickness $\delta_{\text{post-buckl.}}$ can be defined with limit stress being reached with post-buckling behavior which is described by geometrically non-linear correlations. General methodology of load-bearing panels design based on post-buckling state is shown in reference [4].

Thirdly panel maximal thickness $max(\delta_{\text{stab.}}, \delta_{\text{post-buckl.}})$ is defined since normally indicated values are calculated for different loading levels using different criteria.

Let us consider in further details application of design method. Let us transfer equation (6) for linear case for stability task with small flexures $(f \to 0)$:

$$\frac{E\delta^3}{12\left(1-\mu^2\right)}\frac{\pi^4}{4a^3b^3}\left(3n^4a^4+2a^2b^2n^2+3b^4\right)+E\delta\frac{\pi^2n^2b}{2Ra}\left(A_8+\frac{A_9}{2}\right)=\frac{3\pi^2bT_{\text{x}}^{\text{lim}}}{16a},$$

from which we derive analytic equation in panel thickness δ based on stability conditions with limit loading level. In this case it is also necessary to use equation $\partial T_{\text{x}}^{\text{lim}}/\partial n^2=0$ for definition of critical value n_{cr} corresponding to panel buckling.

Further on let us transfer non-linear equation (6) in panel thickness of the following form:

$$\frac{E\delta^3}{12\left(1-\mu^2\right)}\frac{\pi^4}{4a^3b^3}\left(3n^4a^4+2n^2a^2b^2+3b^4\right)+E\delta\frac{\pi^4n^2}{2ab}f^2\left(A_1+A_2+A_3+A_4+A_5+\frac{A_6}{2}+\frac{A_7}{2}\right)+E\delta$$

$$\times\frac{\pi^2b}{2Ra}\left[\left(A_1+\frac{A_2}{2}\right)f+\left(A_8+\frac{A_9}{2}\right)\right]=\frac{3\pi^2\text{b}T_{\text{x}}^{\text{ult}}}{16a}. \quad (9)$$

With action of ultimate loads which differ from limit loads for safety factor. In this case according to design methodology based on post-buckling state [3-4] let us assume that for example with post-buckling behavior longitudinal stress reach limit values in terms of static strength $\sigma_{\text{x}}=\sigma_{\text{B}}$. Let us transfer membrane stress defined by correlations (7) into second degree equation in flexure amplitude f assuming compression stress is positive:

$$\sigma_{\text{x}}=\frac{4\pi^2n^2E}{b^2}f^2\varDelta_1+\frac{4\pi^2n^2Ef}{Rb^2}A_9\varDelta_2+\frac{T_{\text{x}}^{\text{ult}}}{\delta}=\sigma_{\text{B}}.$$

Deriving amplitude f from the last equation and substituting it into equation (9) required non-linear equation for panel best possible thickness with compression flow $T_{\text{x}}^{\text{ult}}$ action can be obtained. Maximum stress in cylindrical panels with mild camber reaches limit values in points $\left(x_{\text{cr}}^{(i)}, y_{\text{cr}}^{(i)}\right)$ defined by magnitude of function $\varDelta_1(x,y)$ (8) which depends only on correlation of panel geometrical parameters a and b.

Now let us consider peculiarities of task of panels design based on post-buckling state taking into account formation of ID (fid $\neq 0$) which depends on taken technological solutions. Let us assume that ID amplitude fid can be defined by expertise as dependence on panel thickness

$$f_{\text{id}}(\delta)=\varOmega(\delta)=A\delta^2+B\delta+C,$$

with A, B, C being coefficients obtained by tests.

Then substituting ID as function $f_{\text{id}}(\delta)$ into above mentioned method correlations non-linear equation in panel thickness with post-buckling state acceptability in case of limit stress being reached with ID can be obtained. For above mentioned case let us transfer equation (6) of the following form:

$$\frac{E\delta^3}{12\left(1-\mu^2\right)}\frac{\pi^4f}{4a^3b^3}\left(3n^4a^4+2n^2a^2b^2+3b^4\right)+E\delta\frac{\pi^4n^2}{2ab}\left(f+A\delta^2+B\delta+C\right)\left(f^2+2fA\delta^2+B\delta+C\right)$$

$$\times\left(A_1+A_2+A_3+A_4+A_5+\frac{A_6}{2}+\frac{A_7}{2}\right)+E\delta\frac{\pi^2 b}{2Ra}\left[\left(A_1+\frac{A_2}{2}\right)\left(f^2+2f\left(A\delta^2+B\delta\right.\right.\right.$$

$$\left.\left.\left.+C\right)\right)+\left(A_8+\frac{A_9}{2}\right)f\right]=\frac{3\pi^2 bT_{\text{x}}^{\text{ult}}\left(f+A\delta^2+B\delta+C\right)}{16a}, \qquad (10)$$

and equation (7) for longitudinal membrane stress of the following form:

$$\frac{4\pi^2 n^2 E}{b^2}\left(f^2+2f\left(A\delta^2+B\delta+C\right)\right)\Delta_1+\frac{4\pi^2 n^2 Ef}{Rb^2}A_9\Delta_2+\frac{T_{\text{x}}^{\text{ult}}}{\delta}=\sigma_{\text{B}}. \qquad (11)$$

In this case method can be framed similarly to the above-mentioned procedure. It is necessary to derive flexure amplitude *f* from the second-degree equation (11) and substitute it into equation (10). As a result, non-linear equation in panel best possible thickness with limit values being reached in points $\left(x_{\text{cr}}^{(i)}, y_{\text{cr}}^{(i)}\right)$.

4. Conclusion

Engineer method defining best possible thickness of cylindrical panel with mild camber based on post-buckling behavior with limit values being reached. Method can be used at an early design stage for choosing structural and technological solutions in order to support developed aircraft structures minimal weight.

References

[1] Merkulov I and Endogur A 2017 Supersonic aircraft welded section model development taking into account structural and technological scheme *Trudy MAI* **94** 30
[2] Volmir A 1956 *Flexible Plates and Shells* (Moscow: Gostekhizdat)
[3] Mitrofanov O 2002 Engineer design of composite wing panels of minimal weight taking into account skin post-buckling behavior *Aerospace MAI Journal* **9(1)** 34–41
[4] Mitrofanov O 2003 *Engineer Composite Load-Bearing Panels Design Methods* (Moscow: Sputnik+)

IOP Conf. Series: Materials Science and Engineering **684** (2019) 012018 doi:10.1088/1757-899X/684/1/012018

Post-buckling behaviour estimation of thin cylindrical composite shells of non-symmetrical structure in case of compression

O V Mitrofanov

Moscow Aviation Institute (National Research University), 4 Volokolamskoye Highway, Moscow 125993, Russia

e-mail: Oleg1Mitrofanov@yandex.ru

Abstract. The paper deals with the result of analytic solution of the geometrically non-linear task for composite thin-walled cylindrical shell post-buckling behavior in case of axial compression by using Bubnov-Galerkin method. Initial correlations take into account composite structure non-symmetry.

1. Introduction

Let us consider geometrically non-linear task analytic solution for composite cylindrical shell in case of axial compression. Let us use the hypotheses of the composite structures theory by V. Vasiliev [1]. It is shown in reference [1] that consideration of structure non-symmetry influence for flat panels of some composite layouts when defining critical compression force can lead to an error of about ~ 30%. It should be noted that similar influence can be expected in post-buckling behavior estimation tasks for thin composite shells with non-symmetric structure.

2. Main correlations and statement of the problem

Let us put down basic geometrically non-linear correlations. Let us write equation of strain compatibility for composite shells of the following form [2]

$$L_1(F) - L_2(W) + L_5(W) = 0 \tag{1}$$

with $L_1(F) = c_y \dfrac{\partial^4 F}{\partial x^4} + \left(c - 2c_{xy}\right)\dfrac{\partial^4 F}{\partial x^2 \partial y^2} + c_x \dfrac{\partial^4 F}{\partial y^4}$, $L_2(W) = \left(\dfrac{\partial^2 W}{\partial x \partial y}\right)^2 - \left(\dfrac{\partial^2 W}{\partial x^2}\right)\left(\dfrac{\partial^2 W}{\partial y^2}\right) - \dfrac{1}{R}\dfrac{\partial^2 W}{\partial x^2}$,

$$L_5(W) = B_{yx}\frac{\partial^4 W}{\partial x^4} + \left(B_x + B_y - 2C\right)\frac{\partial^4 W}{\partial x^2 \partial y^2} + B_{xy}\frac{\partial^4 W}{\partial y^4}.$$

Let us present Karman non-linear equation of the following form:

$$L_4(W) - L_3(F,W) - L_6(F) = 0$$

with $L_3(F,W) = \frac{\partial^2 F}{\partial y^2}\frac{\partial^2 W}{\partial x^2} + \frac{\partial^2 F}{\partial x^2}\frac{\partial^2 W}{\partial y^2} - 2\frac{\partial^2 F}{\partial x \partial y}\frac{\partial^2 W}{\partial x \partial y} + \frac{1}{R}\frac{\partial^2 F}{\partial x^2}$,

$L_4(W) = D_x \frac{\partial^4 W}{\partial x^4} + \left(4D + D_{xy} + D_{yx}\right)\frac{\partial^4 W}{\partial x^2 \partial y^2} + D_y \frac{\partial^4 W}{\partial y^4}$,

$L_6(F) = C_{xy}\frac{\partial^4 F}{\partial x^4} + 2\left(C_x + C_y - 2C\right)\frac{\partial^4 F}{\partial x^2 \partial y^2} + C_{yx}\frac{\partial^4 F}{\partial y^4}$.

Sandwich shell generalized stiffness is defined by following correlations:

$$B_{mn} = I_{mn}^{(0)},\ C_{mn} = I_{mn}^{(1)} - \mathrm{e} I_{mn}^{(0)},\ D_{mn} = I_{mn}^{(2)} - 2\mathrm{e} I_{mn}^{(1)} + \mathrm{e}^2 I_{mn}^{(0)},$$

$$I_{mn}^{(r)} = \frac{1}{r+1}\sum_{i=1}^{K} A_{mn}^{(i)}\left(t_i^{r+1} - t_{i-1}^{r+1}\right)\ (r = 0, 1, 2),\quad E_{11,22} = \frac{E_{1,2}}{1 - \mu_{12}\mu_{21}},$$

$$A_{11}^{(i)} = \overline{E}_1^{(i)}\cos^4\phi_i + \overline{E}_2^{(i)}\sin^4\phi_i + 2(\overline{E}_1^{(i)}\mu_{12}^{(i)} + 2G_{12}^{(i)})\sin 2\phi_i \cos 2\phi_i,$$

$$A_{12}^{(i)} = A_{21}^{(i)} = \overline{E}_1^{(i)}\mu_{12}^{(i)} + [\overline{E}_1^{(i)} + \overline{E}_2^{(i)} - 2(\overline{E}_1^{(i)}\mu_{12}^{(i)} + 2G_{12}^{(i)})\sin 2\phi_i \cos 2\phi_i],$$

$$A_{22}^{(i)} = \overline{E}_1^{(i)}\sin^4\phi_i + \overline{E}_2^{(i)}\cos^4\phi_i + 2(\overline{E}_1^{(i)}\mu_{12}^{(i)} + 2G_{12}^{(i)})\sin 2\phi_i \cos 2\phi_i,$$

$$A_{33}^{(i)} = [\overline{E}_1^{(i)} + \overline{E}_2^{(i)} - 2\overline{E}_1^{(i)}\mu_{12}^{(i)}]\sin 2\phi_i \cos 2\phi + G_{12}^{(i)}\cos^2 2\phi.$$

Hereinafter following designations are used [1]:

$$c_x = \frac{B_{22}}{B}, c_y = \frac{B_{11}}{B}, c_{xy} = c_{yx} = \frac{B_{12}}{B}, c = \frac{1}{B_{33}},\ B = B_{11}B_{22} - B_{12}^2,$$

$$B_x = \frac{1}{B}(B_{22}C_{11} - B_{12}C_{12}), B_y = \frac{1}{B}(B_{11}C_{22} - B_{12}C_{12}), B_{xy} = \frac{1}{B}(B_{22}C_{12} - B_{12}C_{22}),$$

$$B_{yx} = \frac{1}{B}(B_{11}C_{12} - B_{12}C_{11}), C = \frac{C_{33}}{B_{33}}, C_x = c_x\left(C_{11} - \frac{c_{yx}}{c_x}C_{12}\right)$$

$$C_y = c_y\left(C_{22} - \frac{c_{xy}}{c_y}C_{12}\right), C_{xy} = c_y\left(C_{12} - \frac{c_{xy}}{c_y}C_{11}\right), C_{yx} = c_x\left(C_{12} - \frac{c_{yx}}{c_x}C_{22}\right),$$

$$D_x = D_{11} - C_{11}B_x - C_{12}B_{yx},\ D_y = D_{22} - C_{22}B_y - C_{12}B_{xy},$$

$$D_{xy} = D_{12} - C_{11}B_{xy} - C_{12}B_y,\ D_{yx} = D_{12} - C_{22}B_{yx} - C_{12}B_x,$$

$$D = \frac{D_{33}B_{33} - C_{33}^2}{B_{33}}.$$

3. Shell post-buckling behavior task analytic solution

Let us represent flexure as

$$W = f_1 \cdot \sin\frac{\pi m x}{L}\sin\frac{n y}{R} + f_2 \sin^2\frac{\pi m x}{L}\sin^2\frac{n y}{R}. \quad (2)$$

Herein the first member corresponds to linear task solution and the second one – to non-linear task solution, m and n are corresponding quantity of half waves in longitudinal and circular direction.

Substituting flexure (2) to equation (1) and using method of undetermined coefficients Airy function can be obtained for case of compression with T_x force of the following form:

$$F = A_1\cos\frac{2\pi mx}{L} + A_2\cos\frac{2ny}{R} + A_3\cos\frac{4\pi mx}{L} + A_4\cos\frac{4ny}{R} + A_5\sin\frac{\pi mx}{L}\sin\frac{ny}{R} + A_6\cos\frac{2\pi mx}{L}\cos\frac{2ny}{R}$$

$$+A_7\cos\frac{4\pi mx}{L}\cos\frac{2ny}{R} + A_8\cos\frac{2\pi mx}{L}\cos\frac{4ny}{R} + A_9\sin\frac{3\pi mx}{L}\sin\frac{ny}{R} + A_{10}\sin\frac{\pi mx}{L}$$

$$\times\sin\frac{3ny}{R} + \frac{T_x y^2}{2}$$

with $A_1 = A_1^{(1)}\left(f_1^2 + f_2^2\right) + A_1^{(2)} f_2$, $A_1^{(1)} = \dfrac{\beta^2}{32\alpha^2 c_y}$, $A_1^{(2)} = -\dfrac{1}{16\alpha^2 c_y R} - \dfrac{B_{yx}}{4c_y}$,

$$A_2 = A_2^{(1)}\left(f_1^2 + f_2^2\right) + A_2^{(2)} f_2,\ A_2^{(1)} = \frac{\alpha^2}{32\beta^2 c_x},\ A_2^{(2)} = -\frac{B_{xy}}{4c_x},\ A_3 = A_3^{(1)} f_2^2,\ A_3^{(1)} = -\frac{\alpha^2}{512\beta^2 c_y},$$

$$A_4 = A_4^{(1)} f_2^2,\ A_4 = -\frac{\beta^2}{512\alpha^2 c_x},\ A_5 = A_5^{(1)} f_1 + A_5^{(2)} f_1 f_2 + A_5^{(3)} f_2,$$

$$A_5^{(1)} = \frac{\alpha^2/R}{c_y\alpha^4 + \left(c - c_{xy}\right)\alpha^2\beta^2 + c_x\beta^4},\ A_5^{(2)} = -\frac{\alpha^2\beta^2}{c_y\alpha^4 + \left(c - c_{xy}\right)\alpha^2\beta^2 + c_x\beta^4},$$

$$A_5^{(3)} = \frac{\left(B_{yx}\alpha^4 + \left(B_x + B_y - 2C\right)\alpha^2\beta^2 + B_{xy}\beta^4\right)}{c_y\alpha^4 + \left(c - c_{xy}\right)\alpha^2\beta^2 + c_x\beta^4},\ A_6 = A_6^{(1)} f_2 + A_6^{(2)} f_2^2,$$

$$A_6^{(1)} = \frac{\alpha^2/R + 4\left(B_{yx}\alpha^4 + \left(B_x + B_y - 2C\right)\alpha^2\beta^2 + B_{xy}\beta^4\right)}{16\left[c_y\alpha^4 + \left(c - c_{xy}\right)\alpha^2\beta^2 + c_x\beta^4\right]}, \tag{3}$$

$$A_6^{(2)} = -\frac{\alpha^2\beta^2}{16\left[c_y\alpha^4 + \left(c - c_{xy}\right)\alpha^2\beta^2 + c_x\beta^4\right]},\ A_7 = A_7^{(1)} f_2^2,$$

$$A_7^{(1)} = \frac{\alpha^2\beta^2}{32\left[16c_y\alpha^4 + 4\left(c - c_{xy}\right)\alpha^2\beta^2 + c_x\beta^4\right]},\ A_8 = A_8^{(1)} f_2^2,$$

$$A_8^{(1)} = \frac{\alpha^2\beta^2}{32\left[c_y\alpha^4 + 4\left(c - c_{xy}\right)\alpha^2\beta^2 + 16c_x\beta^4\right]},\ A_9 = A_9^{(1)} f_1 f_2,$$

$$A_9^{(1)} = \frac{3\alpha^2\beta^2}{2\left[81c_y\alpha^4 + 9\left(c - c_{xy}\right)\alpha^2\beta^2 + c_x\beta^4\right]},\ A_{10} = A_{10}^{(1)} f_1 f_2,$$

$$A_{10}^{(1)} = \frac{3\alpha^2\beta^2}{2\left[c_y\alpha^4 + 9\left(c - c_{xy}\right)\alpha^2\beta^2 + 81c_x\beta^4\right]}.$$

Membrane stress at the moment of shell buckling can be derived from stress function definition:

$$\sigma_{\mathrm{x}} = \frac{1}{\delta}\frac{\partial^2 F}{\partial y^2},\ \sigma_{\mathrm{y}} = \frac{1}{\delta}\frac{\partial^2 F}{\partial x^2},\ \tau_{\mathrm{xy}} = \frac{1}{\delta}\frac{\partial^2 F}{\partial x \partial y}. \tag{4}$$

According to Bubnov-Galerkin method procedure the following equations are obtained:

$$\int_0^{2\pi R}\int_0^{L}\left[L_4(W) - L_4(F,W) - L_6(F)\right]\sin\frac{\pi m x}{L}\sin\frac{n y}{R}\mathrm{d}x\,\mathrm{d}y = 0,$$

$$\int_0^{2\pi R}\int_0^{L}\left[L_4(W) - L_4(F,W) - L_6(F)\right]\sin^2\frac{\pi m x}{L}\sin^2\frac{n y}{R}\mathrm{d}x\,\mathrm{d}y = 0.$$

Further with $f_1 \neq 0$ and $f_2 \neq 0$ following non-linear equation system correlating flexure amplitudes and acting compression force can be obtained after lengthy transformations:

$$\begin{aligned}
&\frac{\pi RL}{2} f_1\left[D_{\mathrm{x}}\alpha^4 + \left(4D + D_{\mathrm{xy}} + D_{\mathrm{yx}}\right)\alpha^2\beta^2 + D_{\mathrm{y}}\beta^4\right] - \frac{\pi RL}{2}\alpha^2 f_1 T_{\mathrm{x}} + \alpha^2\beta^2\pi RL\Big[-f_1\Big(A_2^{(1)}\left(f_1^2 + f_2^2\right) \\
&\quad + A_2^{(2)} f\Big) + \frac{3}{8} f_2\left(A_5^{(1)} f_1 + A_5^{(2)} f_1 f_2 + A_5^{(3)} f_2\right) + \frac{1}{2} f_1\left(A_6^{(1)} f_2 + A_6^{(2)} f_2^2\right) - \frac{3}{8} f_1 f_2^2 A_9^{(1)} \\
&\quad - \frac{9}{8} f_1 f_2^2 A_{10}^{(1)}\Big] + \alpha^2\beta^2\pi RL\Big[-f_1\left(A_1^{(1)}\left(f_1^2 + f_2^2\right) + A_1^{(2)} f_2\right) + \frac{3}{8} f_2\Big(A_5^{(1)} f_1 + A_5^{(2)} f_1 f_2 \\
&\quad + A_5^{(3)} f_2\Big) - \frac{9}{8} f_1 f_2^2 A_9^{(1)} - \frac{3}{8} f_1 f_2^2 A_{10}^{(1)}\Big] + \alpha^2\beta^2\pi RL\Big[\frac{1}{4} f_2\left(A_5^{(1)} f_1 + A_5^{(2)} f_1 f_2 + A_5^{(3)} f_2\right) \\
&\quad - f_1\left(A_6^{(1)} f_2 + A_6^{(2)} f_2^2\right) - \frac{3}{8} f_1 f_2^2 A_9^{(1)} - \frac{3}{8} f_1 f_2^2 A_{10}^{(1)}\Big] - \frac{\alpha^2 \pi LR}{2}\Big(A_5^{(1)} f_1 + A_5^{(2)} f_1 f_2 \\
&\quad + A_5^{(3)} f_2\Big) + \pi RL\frac{\left(A_5^{(1)} f_1 + A_5^{(2)} f_1 f_2 + A_5^{(3)} f_2\right)}{2}\Big[C_{\mathrm{xy}}\alpha^4 + 2\left(C_{\mathrm{x}} + C_{\mathrm{y}} - 2C\right)\alpha^2\beta^2 \\
&\quad + C_{\mathrm{yx}}\beta^4\Big] = 0,
\end{aligned} \tag{5}$$

$$\begin{aligned}
&\frac{\pi RL}{2} f_2\left[3D_{\mathrm{x}}\alpha^4 + \left(4D + D_{\mathrm{xy}} + D_{\mathrm{yx}}\right)\alpha^2\beta^2 + 3D_{\mathrm{y}}\beta^4\right] - \frac{3\pi RL}{8}\alpha^2 f_2 T_{\mathrm{x}} + \alpha^2\beta^2\pi RL\Big[-f_1\Big(A_2^{(1)}\Big(f_1^2 \\
&\quad + f_2^2\Big) + A_2^{(2)} f_2\Big) + A_4^{(1)} f_2^3 + \frac{9 f_1}{32}\left(A_5^{(1)} f_1 + A_5^{(2)} f_1 f_2 + A_5^{(3)} f_2\right) + f_2\left(A_6^{(1)} f_2 + A_6^{(2)} f_2^2\right) \\
&\quad - \frac{1}{2} f_2^3 A_7^{(1)} - f_2^3 A_8^{(1)} - \frac{3}{32} f_1^2 f_2 A_9^{(1)} - \frac{27}{32} f_1^2 f_2 A_{10}^{(1)}\Big] + \alpha^2\beta^2\pi RL\Big[-f_2\Big(A_1^{(1)}\left(f_1^2 + f_2^2\right) \\
&\quad + A_1^{(2)} f_2\Big) + A_3^{(1)} f_2^3 + \frac{9}{32} f_1\left(A_5^{(1)} f_1 + A_5^{(2)} f_1 f_2 + A_5^{(3)} f_2\right) + f_2\left(A_6^{(1)} f_2 + A_6^{(2)} f_2^2\right) - \\
&\quad - f_2^3 A_7^{(1)} - \frac{1}{2} f_2^3 A_8^{(1)} - \frac{27}{32} f_1^2 f_2 A_9^{(1)} - \frac{9}{32} f_1^2 f_2 A_{10}^{(1)}\Big] - 2\alpha^2\beta^2\pi RL\Big[\frac{1}{32} f_2 + \Big(A_5^{(1)} f_1 \\
&\quad + A_5^{(2)} f_1 f_2 + A_5^{(3)} f_2\Big) + \frac{1}{2} f_1\left(A_6^{(1)} f_2 + A_6^{(2)} f_2^2\right) - \frac{1}{2} f_2^3 A_7^{(1)} - \frac{1}{2} f_2^3 A_8^{(1)} - \frac{3}{32} f_1^2 f_2 A_9^{(1)}
\end{aligned}$$

$$-\frac{3}{32}f_1^2 f_2 A_{10}^{(1)}\Big] - \frac{\alpha^2 \pi LR}{2}\left(A_6^{(1)} f_2 + A_6^{(2)} f_2^2\right) + \pi RL\Big[2C_{xy}\alpha^4\Big(A_6^{(1)} f_2 + A_6^{(2)} f_2^2$$

$$-2A_1^{(1)}\left(f_1^2 + f_2^2\right) - 2A_1^{(2)} f_2\Big) + 2\left(C_x + C_y - 2C\right)\alpha^2\beta^2\left(A_6^{(1)} f_2 + A_6^{(2)} f_2^2\right)$$

$$+C_{yx}\beta^4\left(4A_2^{(1)}\left(f_1^2 + f_2^2\right) + 4A_2^{(2)} f_2 + 2A_6^{(1)} f_2 + 2A_6^{(2)} f_2^2\right)\Big] = 0. \quad (6)$$

Shell mode of deformation with post-buckling behavior can be defined by the following procedure. At first, when putting down equation system (5)-(6) for linear case it is necessary to consider stability task and calculate critical parameters of wave generation m_{cr} and n_{cr} when dropping non-linear members from $f_1^k \to 0$ (with $k \geq 2$) and members from $f_2^k \to 0$ (responsible for non-linear behavior). Secondly, non-linear equation system (5)-(6) shall be solved taking into account obtained half waves quantity relative to flexure amplitude f_1 and f_2 . In this case strain state with shell post-buckling behavior can be defined based on the known flexure function (2). Stress with non-linear behavior after buckling can be defined by correlations (4). Equation for longitudinal stress is shown herein as an example:

$$\sigma_x = \frac{n^2}{\delta R}\Big\{4A_2\cos\frac{2ny}{R} + 16A_4\cos\frac{4ny}{R} + A_5\sin\frac{\pi mx}{L}\sin\frac{ny}{R} + 4A_6\cos\frac{2\pi mx}{L}\cos\frac{2ny}{R} + 4A_7\cos\frac{4\pi mx}{L}$$

$$\times\cos\frac{2ny}{R} + 16A_8\cos\frac{2\pi mx}{L}\cos\frac{4ny}{R} + A_9\sin\frac{3\pi mx}{L}\sin\frac{ny}{R} + 9A_{10}\sin\frac{\pi mx}{L}\sin\frac{3ny}{R}\Big\} + \frac{T_x}{\delta}.$$

It should be noted that coefficients A_k depend on flexure amplitude (ref. designations (3)).

4. Conclusion

Geometrically non-linear task analytic solution for compressed composite shell taking into account possible structure non-symmetry is shown. Reduced to non-linear equation system solution in flexure amplitude the task can be used at an early stage of design for taken design solutions efficiency estimation.

References

[1] Vasiliev V 1988 *Composite Structures Mechanics* (Moscow: Mashinostroenie)
[2] Mitrofanov O 2017 Main correlations for post-buckling behavior investigation of load-bearing composite panels of non-symmetric structure taking into account initial deflection with compression *Aktualnye problemy sovremennoi nauki* **2** 186–9

IOP Conf. Series: Materials Science and Engineering **684** (2019) 012019 doi:10.1088/1757-899X/684/1/012019

Design of cylindrical composite panels with mild camber for biaxial compression taking into account geometrical non-linearity with uniform heating and rigid support

O V Mitrofanov and M V Klesareva

Moscow Aviation Institute (National Research University), 4 Volokolamskoye Highway, Moscow 125993, Russia

e-mail: Oleg1Mitrofanov@yandex.ru

Abstract. Based on geometrically non-linear task analytic solution for cylindrical composite panel with mild camber method of best possible design for post-buckling state with biaxial compression is suggested. Uniform heating and panel all-boundary rigid support are taken into account in basic correlations.

1. Introduction

Let us consider fuselage panel best possible thickness definition with acceptable post-buckling behavior in case of loads close to ultimate level. It should be noted that with loading below limit level buckling of the considered panel is not acceptable. Let us assume that the investigation object is cylindrical composite panel with mild camber under biaxial compression with all-boundary rigid support. Uniform heating influence on the panel which can be for example a fuselage skin fragment should also be taken into account. Task of design based on post-buckling state with compression taking into account uniform thermal effect t is of practical interest.

2. Main correlations

Let us use correlations of composite thin-walled structures theory by V. Vasiliev [1]. Let us put down physical equations for composite multiple sandwich taking into account thermal effect:

$$\varepsilon_{\mathrm{x}} = \frac{p_{\mathrm{x}}}{E_{\mathrm{x}}} - \mu_{\mathrm{xy}} \frac{p_{\mathrm{y}}}{E_{\mathrm{y}}} - \alpha_{\mathrm{x}} t , \; \varepsilon_{\mathrm{y}} = \frac{p_{\mathrm{y}}}{E_{\mathrm{y}}} - \mu_{\mathrm{yx}} \frac{p_{\mathrm{x}}}{E_{\mathrm{x}}} - \alpha_{\mathrm{y}} t$$

with $\alpha_{\mathrm{x}}, \alpha_{\mathrm{y}}$ being effective coefficients of linear thermal expansion

$$\alpha_{\mathrm{x}} = \frac{B_{1\mathrm{t}} B_{22} - B_{2\mathrm{t}} B_{12}}{B_{11} B_{22} - B_{12}^2} , \; \alpha_{\mathrm{y}} = \frac{B_{2\mathrm{t}} B_{11} - B_{1\mathrm{t}} B_{12}}{B_{11} B_{22} - B_{12}^2} , \; B_{\mathrm{t}p} = \sum_{i=1}^{k} \overline{h}_i b_{\mathrm{t}p}^i , \, p = 1.2,$$

$$b_{\mathrm{t}1}^i = \overline{E}_1^i \left(\alpha_1^i + \mu_{12}^i \alpha_2^i \right) \cos^2 \phi_i + \overline{E}_2^i \left(\alpha_2^i + \mu_{21}^i \alpha_1^i \right) \sin^2 \phi_i ,$$

$$b_{t2}^{i}=\overline{E}_{1}^{i}\left(\alpha_{1}^{i}+\mu_{12}^{i}\alpha_{2}^{i}\right)\sin^{2}\phi_{i}+\overline{E}_{2}^{i}\left(\alpha_{2}^{i}+\mu_{21}^{i}\alpha_{1}^{i}\right)\cos^{2}\phi_{i}.$$

Considering main correlations of orthotropic body elasticity theory for plane stress condition and solving them in stress following can be derived for the i-th layer with fiber angle ϕ_i:

$$\sigma_{1}^{(i)}=\overline{E}_{1}^{(i)}\left\{\left[\frac{1}{E_{\text{x}}}\left(\cos^{2}\phi_{i}+\mu_{12}^{(i)}\sin^{2}\phi_{i}\right)-\frac{\mu_{\text{yx}}}{E_{\text{x}}}\left(\sin^{2}\phi_{i}+\mu_{12}^{i}\cos^{2}\phi_{i}\right)\right]\sigma_{\text{x}}+\left[-\frac{\mu_{\text{xy}}}{E_{\text{y}}}\left(\cos^{2}\phi_{i}+\mu_{12}^{i}\,sin^{2}\,\phi_{i}\right)\right.\right.$$
$$\left.+\frac{1}{E_{\text{y}}}\left(\sin^{2}\phi_{i}+\mu_{12}^{i}\cos^{2}\phi_{i}\right)\right]\sigma_{\text{y}}+\left[\frac{1}{G_{\text{xy}}}\left(1-\mu_{12}^{i}\right)\sin\phi_{i}\cos\phi_{i}\right]\tau_{\text{xy}}+\left[\alpha_{\text{x}}\left(\cos^{2}\phi_{i}+\mu_{12}^{i}\right.\right.$$
$$\left.\left.\left.\times\sin^{2}\phi_{i}\right)+\alpha_{\text{y}}\left(\sin^{2}\phi_{i}+\mu_{12}^{i}\cos^{2}\phi_{i}\right)\right]\left(\alpha_{1}^{i}+\mu_{12}^{i}\alpha_{2}^{i}\right)\cdot t\right\},\tag{1}$$

$$\sigma_{2}^{(i)}=\overline{E}_{2}^{(i)}\left\{\left[\frac{1}{E_{\text{x}}}\left(\sin^{2}\phi_{i}+\mu_{21}^{(i)}\cos^{2}\phi_{i}\right)-\frac{\mu_{\text{yx}}}{E_{\text{x}}}\left(\cos^{2}\phi_{i}++\mu_{12}^{i}\sin^{2}\phi_{i}\right)\right]\sigma_{\text{x}}+\left[-\frac{\mu_{\text{xy}}}{E_{\text{y}}}\left(\sin^{2}\phi_{i}+\mu_{21}^{i}\right.\right.\right.$$
$$\left.\left.\times\cos^{2}\phi_{i}\right)+\frac{1}{E_{\text{y}}}\left(\cos^{2}\phi_{i}+\mu_{21}^{i}\sin^{2}\phi_{i}\right)\right]\sigma_{\text{y}}+\left[-\frac{1}{G_{\text{xy}}}\left(1-\mu_{12}^{i}\right)\sin\phi_{i}\cos\phi_{i}\right]\tau_{\text{xy}}+\left[\alpha_{\text{x}}\right.$$
$$\left.\left.\times\left(\sin^{2}\phi_{i}+\mu_{12}^{i}\cos^{2}\phi_{i}\right)+\alpha_{\text{y}}\left(\cos^{2}\phi_{i}+\mu_{21}^{i}\sin^{2}\phi_{i}\right)\right]\left(\alpha_{2}^{i}+\mu_{21}^{i}\alpha_{1}^{i}\right)\cdot t\right\},$$

$$\tau_{12}^{i}=G_{12}^{i}\left\{\sin 2\phi_{i}\left[-\frac{1}{E_{\text{x}}}-\frac{\mu_{\text{xy}}}{E_{\text{x}}}\right]\sigma_{\text{x}}+\sin 2\phi_{i}\left[\frac{1}{E_{\text{y}}}+\frac{\mu_{\text{xy}}}{E_{\text{y}}}\right]\sigma_{\text{y}}+\frac{\cos 2\phi_{i}}{G_{\text{xy}}}\tau_{\text{xy}}+\sin 2\phi_{i}\left(\alpha_{\text{y}}-\alpha_{\text{x}}\right)\cdot t\right\}$$

with $\sigma_1^{(i)}$ $\left(E_1^{(i)}\right)$ being normal stress (elasticity modulus) along fibers in the layer with fiber angle $\phi = i$.

Method of thin-walled structures design based on post-buckling state is shown in reference [2]. Let us write non-linear equation of strain compatibility of the following form taking into account cylindrical panel mild camber

$$L_1(F) - L_2(W) = 0,\tag{2}$$

with F being stress function, L_m being functional

$$L_1(F)=\frac{1}{E_{\text{y}}}\frac{\partial^4 F}{\partial x^4}+\left(\frac{1}{G_{\text{xy}}}-\frac{2\mu_{\text{xy}}}{E_{\text{x}}}\right)\frac{\partial^4 F}{\partial x^2\partial y^2}+\frac{1}{E_{\text{x}}}\frac{\partial^4 F}{\partial y^4},\quad L_2(W)=\left(\frac{\partial^2 W}{\partial x\partial y}\right)^2-\left(\frac{\partial^2 W}{\partial x^2}\right)\left(\frac{\partial^2 W}{\partial y^2}\right)-\frac{1}{R}\frac{\partial^2 W}{\partial x^2}.$$

Karman non-linear equation of type [2]

$$L_3(F,W) - L_4(W) = 0\tag{3}$$

with $L_3(F,W)=\frac{\partial^2 F}{\partial y^2}\frac{\partial^2 W}{\partial x^2}+\frac{\partial^2 F}{\partial x^2}\frac{\partial^2 W}{\partial y^2}-2\frac{\partial^2 F}{\partial x\partial y}\frac{\partial^2 W}{\partial x\partial y}+\frac{1}{R}\frac{\partial^2 F}{\partial x^2}$,

$$L_4(W)=\frac{1}{\delta}\left[D_{\mathrm{x}}\frac{\partial^4 W}{\partial x^4}+2D_3\frac{\partial^4 W}{\partial x^2\partial y^2}+D_{\mathrm{y}}\frac{\partial^4 W}{\partial y^4}\right].$$

Equations (2) and (3) are basic geometrically non-linear correlations of composite panel post-buckling behavior task.

3. Non-linear task solution and panels design method

Let us represent panel flexure as

$$W=f\sin^2\alpha x\cdot\sin^2\beta y, \quad (4)$$

with $\alpha=\pi m/a$, $\beta=\pi n/b$ being wave generation parameters.

Substituting flexure (4) into equations (2) stress function of following form can be obtained after some transformations and coefficients definition:

$$F=f^2\delta\{A_1\cos 2\alpha x-A_2\cos 4\alpha+A_3\cos 2\beta yx-A_4\cos 4\beta y-A_5\cos 2\alpha x\cos 2\beta y+A_6\cos 4\alpha x\cos 2\beta y$$
$$+A_7\cos 2\alpha x\cos 4\beta y\}+\frac{f\delta}{R}\{-A_8\cos 2\alpha x+A_9\cos 2\alpha x\cos 2\beta y\}+\frac{N_{\mathrm{x}}y^2}{2}+\frac{N_{\mathrm{y}}x^2}{2}.$$

With following designations introduced:

$$A_1=E_{\mathrm{y}}\frac{\beta^2}{32\alpha^2},\ A_2=\frac{E_{\mathrm{y}}}{512}\frac{\beta^2}{\alpha^2},\ A_3=E_{\mathrm{x}}\frac{\alpha^2}{32\beta^2},\ A_4=\frac{E_{\mathrm{x}}}{512}\frac{\alpha^2}{\beta^2},$$

$$A_5=2\frac{\alpha^2\beta^2}{G_{\alpha\beta}},\ A_6=\frac{\alpha^2\beta^2}{64G_{4\alpha}},\ A_7=\frac{\alpha^2\beta^2}{64G_{4\beta}},\ A_8=\frac{E_{\mathrm{y}}}{16\alpha^2},\ A_9=\frac{1}{16\alpha^2G_{\alpha\beta}},$$

$$G_{\alpha\beta}=\frac{\alpha^4}{E_{\mathrm{y}}}+\left(\frac{1}{G_{\mathrm{xy}}}-\frac{2\mu_{\mathrm{xy}}}{E_{\mathrm{x}}}\right)\alpha^2\beta^2+\frac{\beta^4}{E_{\mathrm{x}}},\ G_{4\alpha}=\frac{16\alpha^4}{E_{\mathrm{y}}}+\left(\frac{1}{G_{\mathrm{xy}}}-\frac{2\mu_{\mathrm{xy}}}{E_{\mathrm{x}}}\right)4\alpha^2\beta^2+\frac{\beta^4}{E_{\mathrm{x}}},$$

$$G_{4\beta}=\frac{\alpha^4}{E_{\mathrm{y}}}+\left(\frac{1}{G_{\mathrm{xy}}}-\frac{2\mu_{\mathrm{xy}}}{E_{\mathrm{x}}}\right)4\alpha^2\beta^2+\frac{16\beta^4}{E_{\mathrm{x}}}.$$

Task solution by Bubnov-Galerkin method is reduced to following equation:

$$\delta^3\overline{D}_{\mathrm{T}}+f^2\delta E_{m1}+\delta E_{m2}-f\delta E_{m3}-\frac{N_{\mathrm{x}}\alpha^2}{2}-\frac{N_{\mathrm{y}}\beta^2}{2}=0 \quad (5)$$

with $\overline{D}_{mn}=\frac{1}{4}\left[3\overline{D}_{\mathrm{x}}\alpha^4+2\overline{D}_3\alpha^2\beta^2+3\overline{D}_{\mathrm{y}}\beta^4\right]$, $\overline{D}_k=\frac{D_k}{\delta^3}$,

$$E_{m1}=\frac{1}{128}\left[\frac{30}{16}E_{\mathrm{x}}\alpha^4+\frac{4\alpha^6\beta^2}{G_{\alpha\beta}}-\frac{\alpha^4\beta^4}{G_{4\alpha}}+\frac{\alpha^4\beta^4}{G_{4\beta}}+\frac{17}{8}E_{\mathrm{y}}\beta^4\right],$$

$$E_{m2}=\frac{1}{32R^2}\left[E_{\mathrm{y}}+\frac{\alpha^4}{2G_{\alpha\beta}}\right],\ E_{m3}=\frac{\beta^2}{4R}\left[\frac{E_{\mathrm{y}}}{16}-\frac{9\alpha^4}{64G_{\alpha\beta}}\right].$$

When considering small flexures in longitudinally compressed cylindrical panel stability task solution following equation in compression force can be obtained:

$$N_x = \frac{\delta^3}{\alpha^2}\left[4D_x\alpha^4 + \frac{8}{3}D_3\alpha^2\beta^2 + 4\overline{D}_y\beta^4\right] + \frac{\delta}{6R^2}\left(\frac{E_y}{\alpha^2} + \frac{\alpha^2}{2\left(\frac{\alpha^4}{E_y} + \left(\frac{1}{G_{xy}} - \frac{2\mu_{xy}}{E_x}\right)\alpha^2\beta^2 + \frac{\beta^4}{E_x}\right)}\right). \quad (6)$$

It should be noted that wave generation critical parameters m_{cr} and n_{cr} are defined from correlations $\partial N_x/\partial m = 0$ and $\partial N_x/\partial n = 0$. Besides equation for combined loading similar to equation (6) can be obtained with fixed correlation of considered forces.

Let us write further formulas for membrane stress in middle surface:

$$\sigma_x = -f^2 S_1 - \frac{f}{R}S_2 - \frac{N_x}{\delta}, \quad (7)$$

$$\sigma_y = -f^2 S_3 - \frac{f}{R}S_4 - \frac{N_y}{\delta}, \quad (8)$$

With following functions designated:

$$S_1 = -4\{\beta^2 A_3\cos 2\beta yx - 4\beta^2 A_4\cos 4\beta y - \beta^2 A_5\cos 2\alpha x\cos 2\beta y + \beta^2 A_6\cos 4\alpha x\cos 2\beta y + 4\beta^2 A_7 \times\cos 2\alpha x\cos 4\beta y\},$$

$$S_2 = 4\beta^2 A_9\cos 2\alpha x\cos 2\beta y,$$

$$S_3 = -4\{\alpha^2 A_1\cos 2\alpha x - 4\alpha^2 A_2\cos 4\alpha - \alpha^2 A_5\cos 2\alpha x\cos 2\beta y + 4\alpha^2 A_6\cos 4\alpha x\cos 2\beta y + \alpha^2 A_7 \times\cos 2\alpha x\cos 4\beta y\},$$

$$S_4 = -4\alpha^2 A_8\cos 2\alpha x + 4\alpha^2 A_9\cos 2\alpha x\cos 2\beta y.$$

Let us state a remark concerning functions $S_i(x, y)$. It should be noted that above mentioned functions are periodic and for definition of (x_j, y_j) coordinates in which maximum stress is observed in general parametric studies are needed. It should also be noted that surfaces $S_i(x, y)$ do not depend on panel thickness and are defined by stiffness parameters (of composite layout) correlations and panel geometrical parameters ratios.

Let us transform equation (1) for stress along fibers in standard layout 0°/±45°/90°. We obtain following:

$$\sigma_1^{(0)} = \overline{E}_1^{(0)}\left\{\frac{\sigma_x}{E_x}\left(1-\mu_{12}^{(0)}\mu_{yx}\right) + \frac{\sigma_y}{E_y}\left(\mu_{12}^{(0)} - \mu_{yx}\right)\right\} + \overline{E}_1^{(0)}\left\{\alpha_x + \alpha_y\mu_{12}^{(0)} - \left(\alpha_1^{(0)} + \mu_{12}^{(0)}\alpha_2^{(0)}\right)\right\}\cdot t, \quad (9)$$

$$\sigma_1^{(90)} = \overline{E}_1^{(90)}\left\{\frac{\sigma_y}{E_y}\left(1-\mu_{12}^{(90)}\mu_{yx}\right) + \frac{\sigma_x}{E_x}\left(\mu_{12}^{(90)} - \mu_{xy}\right)\right\} + \overline{E}_1^{(90)}\left\{\alpha_x\mu_{12}^{(90)} + \alpha_y - \left(\alpha_1^{(90)} + \mu_{12}^{(90)} \times\alpha_2^{(90)}\right)\right\}\cdot t \quad (10)$$

$$\sigma_1^{(\pm45)} = \frac{\overline{E}_1^{(\pm45)}}{2}\left\{\frac{\sigma_x}{E_x}\left(1-\mu_{yx}\right)\left(1+\mu_{12}^{(\pm45)}\right)+\frac{\sigma_y}{E_y}\left(1-\mu_{xy}\right)\left(1+\mu_{12}^{(\pm45)}\right)\right\}+\frac{\overline{E}_1^{(\pm45)}}{2}\left\{\alpha_x\left(1+\mu_{12}^{(\pm45)}\right)+\alpha_y\right.$$

$$\left.\times\left(1+\mu_{12}^{(\pm45)}\right)+\alpha_{xy}\left(1-\mu_{12}^{(\pm45)}\right)-\left(\alpha_1^{(\pm45)}+\mu_{12}^{(\pm45)}\alpha_2^{(\pm45)}\right)\right\}\cdot t. \quad (11)$$

Shown equations (9)–(11) can be used for composite package stress limit definition according to the first structure theory for combined loading with acting stress σ_x and σ_y. In this case with stress along fibers tendency to ultimate stress limit $\left(\sigma_1^{(i)} \to \overline{\sigma}_1^{(i)}\right)$ irreducible combined load can be defined.

Assuming that possible reason of standard composite package 0°/±45°/90° failure is failure of layer fibers with reinforcement $\varphi = 0°$. Then substituting equations (7)–(8) which describe post-buckling behavior into equation (9) we derive

$$\overline{\sigma}_1^{(0)} = \overline{E}_1^{(0)}\frac{1-\mu_{12}^{(0)}\mu_{yx}}{E_x}\left(-f^2S_1-\frac{f}{R}S_2-\frac{T_x}{\delta}\right)+\overline{E}_1^{(0)}\frac{\mu_{12}^{(0)}-\mu_{yx}}{E_y}\left(-f^2S_3-\frac{f}{R}S_4-\frac{T_y}{\delta}\right)+\overline{E}_1^{(0)}\left\{\alpha_x\right.$$

$$\left.+\alpha_y\mu_{12}^{(0)}-\left(\alpha_1^{(0)}+\mu_{12}^{(0)}\alpha_2^{(0)}\right)\right\}\cdot t.$$

From which with given loads per unit $T_x = \sigma_x\delta$ and $T_y = \sigma_y\delta$ following is derived

$$f = \frac{-B+\sqrt{B^2-4AC}}{2A}, \quad (12)$$

with $A = \overline{E}_1^{(0)}\frac{1-\mu_{12}^{(0)}\mu_{yx}}{E_x}S_1+\overline{E}_1^{(0)}\frac{\mu_{12}^{(0)}-\mu_{yx}}{E_y}S_3$, $B = \overline{E}_1^{(0)}\frac{1-\mu_{12}^{(0)}\mu_{yx}}{RE_x}S_2+\overline{E}_1^{(0)}\frac{\mu_{12}^{(0)}-\mu_{yx}}{RE_y}S_4$,

$$C = \overline{E}_1^{(0)}\frac{1-\mu_{12}^{(0)}\mu_{yx}}{E_x}\frac{T_x}{\delta}+\overline{E}_1^{(0)}\frac{\mu_{12}^{(0)}-\mu_{yx}}{RE_y}\frac{T_y}{\delta}+\left\{\alpha_x+\alpha_y\mu_{12}^{(0)}-\left(\alpha_1^{(0)}+\mu_{12}^{(0)}\alpha_2^{(0)}\right)\right\}\cdot t.$$

Further substituting (12) into (5) let us put down equation defining cylindrical panel composite package minimal thickness in case of breaking composite layer fibers with longitudinal reinforcement

$$\delta^3\overline{D}_T+\left(\frac{-B+\sqrt{B^2-4AC}}{2A}\right)^2\delta E_{m1}+\delta E_{m2}-\sqrt{\left(\frac{-B+\sqrt{B^2-4AC}}{2A}\right)}\delta E_{m3}-\frac{N_x\alpha^2}{2}-\frac{N_y\beta^2}{2}=0. \quad (13)$$

Applying transformations with possible fibers breaking in layers with reinforcement $\varphi = \pm45°$ and $\varphi = 90°$ similar non-linear equations in thickness can be obtained providing fibers breakage in a layer with one of the above-mentioned fiber angles according to the 1-st structure theory. Thus cylindrical composite panels design method based on post-buckling state with biaxial compression taking into account thermal affect is reduced to following. At first thicknesses are calculated using equations of type (13) with possible failure reasons being fiber breakage with reinforcement 0°, ±45° and 90°. Further maximum value which is the target thickness δ post-buck of cylindrical composite panel is chosen.

It should be noted that shown method can be appended with consideration of stability limitations. That is loads used for suggested method should be ultimate ones. Generally for composite structures buckling is not acceptable within limit loading level. In this case with given load value and known composite layout from equation (6) panel thickness δ_{stab} can be defined and further maximum thickness value should be chosen (δ_{stab}, $\delta_{post\text{-}buck}$).

4. Conclusion

Shown design method for cylindrical composite panels with mild camber based on post-buckling state with biaxial compression taking into account uniform heating and rigid support can be used for aircraft structures design with limitations in stability and load-bearing capacity.

References

[1] Vasiliev V 1988 *Composite Structures Mechanics* (Moscow: Mashinostroenie)
[2] Mitrofanov O 2003 *Engineer Composite Load-Bearing Panels Design Methods* (Moscow: Sputnik+)

Design of rectangular sandwich panels with metal skins based on post-buckling state in compression and shear

O V Mitrofanov

Moscow Aviation Institute (National Research University), 4 Volokolamskoye Highway, Moscow 125993, Russia

e-mail: Oleg1Mitrofanov@yandex.ru

Abstract. Solutions of geometrically non-linear tasks concerning post-buckling behavior of sandwich panels with thin metal skins and lightweight filler loaded by compression and shear flows are shown. Based on obtained solutions methods of aircraft structures sandwich panels design based on post-buckling state with acceptable global buckling in case of ultimate loading are suggested.

1. Introduction

Sandwich panels (SP) bending stiffness is significant, they are widely used for aircraft structures. SP usually consist of thin load-bearing skins and lightweight filler layer. Buckling is not allowed with loads under limit level in SP design. But buckling is acceptable with loads close to ultimate level. Moreover, tasks concerning SP load-bearing capacity are of interest for experimental determination of aircraft structures actual safety margins when testing them up to failure. It should be noted that SP best possible design tasks are actual for minimal weight aircraft structures development. With that both types of panels can be considered: panels of main load-bearing structures and of secondary elements such as fairings of different purpose.

Let us consider further tasks of SP post-buckling behavior with loads above limit level under compression and tangential flows. Then methods for SP parameters definition with acceptable post-buckling behavior for ultimate loading and limit stress $\bar{\sigma}$ are presented. It should be noted that limit stress can be understood to be either yield stress $\bar{\sigma}=\sigma_{\mathrm{y}}$ or ultimate stress limit $\bar{\sigma}=\sigma_{\mathrm{u}}$ depending on structural peculiarities of the panel when designing aircraft structures. It should be noted that in this investigation only geometrical non-linearity of the task but not its physical non-linearity will be taken into account. In order to take into account influence of loose aggregate basic Reisner correlations [1] are used further on. Method suggested in investigation [2] will be used as basic one for design based on post-buckling state.

2. Main correlations

Let us put down main geometrically non-linear correlations for SP [1] of the following form

$$\frac{1}{E\delta}\nabla^2\nabla^2 F=\left(\frac{\partial^2 W}{\partial x\partial y}\right)^2-\frac{\partial^2 W}{\partial x^2}\frac{\partial^2 W}{\partial y^2}, \tag{1}$$

$$D\nabla^2\nabla^2 W = \left(1-\frac{D}{hG_c}\nabla^2\right)L(W,F).$$

With functional

$$L(W,F)=\frac{\partial^2 F}{\partial y^2}\frac{\partial^2 W}{\partial x^2}+\frac{\partial^2 F}{\partial x^2}\frac{\partial^2 W}{\partial y^2}-2\frac{\partial^2 F}{\partial x\partial y}\frac{\partial^2 W}{\partial x\partial y},$$

$$\nabla^2\nabla^2()=\frac{\partial^4()}{\partial x^4}+2\frac{\partial^4()}{\partial x^2\partial yx^2}+\frac{\partial^4()}{\partial y^4}.$$

SP bending stiffness being $D=\dfrac{Eth^2}{2\left(1-\mu^2\right)}$, F being stress function, t being SP single skin thickness ($\delta=2t$), h being distance between skins middle surfaces, G_c being filler shear modulus.

3. Sandwich panels design for compression

Let us examine rectangular panel with geometrical parameters (a, b) and hinge support. Let us further on assume due to structural and technological reasons that known value is filler height and it is necessary to define skin thickness.

Panel flexure is of the following form:

$$W = f\sin\alpha x\sin\beta y, \tag{2}$$

with $\alpha=\dfrac{\pi m}{a}$, $\beta=\dfrac{\pi n}{b}$; m, n being half waves quantity.

After substitution of flexure (2) into equation of strain compatibility (1) Airy stress function can be obtained:

$$F = f^2\left[A_1\cos 2\alpha x + A_2\cos 2\beta y\right]+\frac{T_x y^2}{2} \tag{3}$$

with $A_1=\dfrac{E\delta}{32}\dfrac{\beta^2}{\alpha^2}$, $A_2=\dfrac{E\delta}{32}\dfrac{\alpha^2}{\beta^2}$.

Equation for Bubnov-Galerkin method application procedure is written in the following form:

$$\int_0^a\int_0^b\left\{D\left[\frac{\partial^4 W}{\partial x^4}+2\frac{\partial^4 W}{\partial x^2\partial yx^2}+\frac{\partial^4 W}{\partial y^4}\right]-\left(1-\frac{D}{hG_c}\nabla^2\right)L(W,F)\right\}\sin\alpha x\sin\beta y=0. \tag{4}$$

As a result of substituting flexure (2) and stress function (3) into equation (4) following non-linear equation is derived

$$D\frac{\pi^4\left(a^2n^2+b^2m^2\right)^2}{16a^3b^3}+\frac{\pi^4 f^2m^2n^2\left(A_1+A_2\right)}{2ab}+\frac{D}{hG_c}\frac{\pi^6 f^2m^2n^2}{2a^3b^3}A_{mn}=T_x\frac{\pi^2m^2b}{4a} \tag{5}$$

with $A_{mn}=A_1a^2n^2+5A_1b^2m^2+5A_2a^2n^2+A_2b^2m^2$.

Longitudinal membrane stress due to buckling are written based on stress function definition

$$\sigma_{\text{x}} = \frac{1}{\delta}\frac{\partial^2 F}{\partial y^2} = -\frac{4\pi^2 A_2 f^2 n^2}{\delta b^2}\cos\frac{2\pi n y}{b} - \frac{T_{\text{x}}}{\delta}. \tag{6}$$

Let us examine further on SP design method assuming that normal stress reaches limits σ_{u} with ultimate compression load. Flexure amplitude is derived from the last equation

$$f = \sqrt{\left(\sigma_{\text{u}} - \frac{T_{\text{x}}}{\delta}\right)\frac{\delta b^2}{4\pi^2 n^2 A_2}}. \tag{7}$$

It should be noted that half waves quantity m and n being part of equations (5)–(7) are defined when solving linear stability task for equation (5) with provision $f^2 \to 0$ and minimizing in above mentioned parameters according to known procedures [3] and defining critical values m_{cr} and n_{cr}. After substituting these values and flexure (7) into equation (5) following non-linear equation in skin thickness is derived:

$$\frac{E\delta h^2}{4\left(1-\mu^2\right)}\frac{\pi^4\left(a^2n^2+b^2m^2\right)^2}{16a^3b^3} + \left(\sigma_{\text{u}} - \frac{T_{\text{x}}}{\delta}\right)\frac{\delta b^2}{4\pi^2 n^2 A_2}\left[\frac{\pi^4 m^2 n^2\left(A_1+A_2\right)}{2ab} + \frac{E\delta h^2}{4\left(1-\mu^2\right)}\frac{1}{hG_{\text{c}}}\frac{\pi^6 m^2 n^2}{2a^3b^3}\right.$$

$$\left.\times A_{mn}\right] - T_{\text{x}}^{\text{est}}\frac{\pi^2 m^2 b}{4a} = 0.$$

Which is correlated to given ultimate compression force $T_{\text{x}}^{\text{ult}}$.

4. Sandwich panel design for shear

Let us examine similar SP design task based on post-buckling state with acting tangential force S_{xy}. For rectangular SP flexure is traditionally represented as

$$W = f\sin\frac{\pi y}{b}\sin\frac{\pi(x-\alpha y)}{s} \tag{9}$$

with α being waves angle, s being distance between waves for buckling.

After substituting flexure (9) into equation of strain compatibility (1) stress function can be derived as

$$F = f^2\left[A_1\cos\frac{2\pi y}{b} + A_2\cos\frac{2(x-\alpha y)}{s}\right] + \frac{T_{\text{x}}y^2}{2} + S_{\text{xy}}xy \tag{10}$$

with $A_1 = \frac{E\delta}{32}\frac{b^2}{s^2}$, $A_2 = \frac{E\delta}{32}\frac{s^2}{b^2\left(\alpha^2+1\right)^2}$.

Here is also an equation using Bubnov-Galerkin method procedure for considered case

$$\int_0^a\int_0^b\left\{D\left[\frac{\partial^4 W}{\partial x^4} + 2\frac{\partial^4 W}{\partial x^2 \partial y x^2} + \frac{\partial^4 W}{\partial y^4}\right] - \left(1 - \frac{D}{hG_{\text{c}}}\nabla^2\right)L(W,F)\right\}\sin\frac{\pi y}{b}\sin\frac{2\pi(x-\alpha y)}{s} = 0. \tag{11}$$

When substituting flexure (9) and stress function (10) into equation (11) following non-linear equation for tangential force S_{xy} case is obtained:

$$D\frac{\pi^4}{4s^3b^3}B_{S1}+f^2\left[\frac{\pi^4 f^2\left(A_1+A_2\right)}{2bs}+\frac{D\pi^6}{hG_c 2s^5b^3}B_{S2}\right]=S_{xy}\frac{\pi^2 b}{4s}, \qquad (12)$$

with $B_{S1}=\alpha^4b^4+2\alpha^2b^4+6\alpha^2b^2s^2+b^4+2b^2s^2+s^4$,

$B_{S2}=A_1s^4+A_2s^4-2A_2\alpha^2b^4-4A_2\alpha^4b^4+A_1b^2s^2+5A_2s^2b^2+2A_2\alpha^2b^2s^2$.

Membrane stress due to buckling is written in general terms as

$$\sigma_x=\frac{1}{\delta}\frac{\partial^2 F}{\partial y^2}=-\frac{4\pi^2 f^2}{\delta b^2 s^2}\left[A_1s^2\cos\frac{2\pi n y}{b}+A_2\alpha^2b^2\cos\frac{2\pi\left(x-\alpha y\right)}{s}\right]-\frac{T_x}{\delta},$$

$$\tau_{xy}=-\frac{1}{\delta}\frac{\partial^2 F}{\partial x\partial y}=\frac{4\pi^2A_2\alpha f^2}{s^2}\cos\frac{2\pi\left(x-\alpha y\right)}{s}+\frac{S_{xy}}{\delta}.$$

It should be noted that critical parameters of wave generation are defined when considering linear stability task using provision $f^2\rightarrow 0$ for equation (12).

Let us further on assume that at the moment of SP buckling skin stress reaches limits τ_u. Then following equation for flexure amplitude is derived from the last equation:

$$f=\sqrt{\left(\tau_u-\frac{S_{xy}}{\delta}\right)\frac{\delta s^2}{4\pi^2A_2\alpha\cos\frac{2\pi\left(x-\alpha y\right)}{s}}}.$$

Substituting it into equation (12) following non-linear equation in skin thickness with given load and filler height can be derived:

$$\frac{Eth^2}{2\left(1-\mu^2\right)}\frac{\pi^4}{4s^3b^3}B_{S1}+\left(\tau_u-\frac{S_{xy}}{\delta}\right)\frac{s^2}{4\pi^2A_2\alpha}\left[\frac{\pi^4 f^2\left(A_1+A_2\right)}{2bs}+\frac{Eth^2}{2\left(1-\mu^2\right)}\frac{\pi^6}{hG_c 2s^5b^3}B_{S2}\right]=S_{xy}\frac{\pi^2b}{4s}.$$

At conclusion a remark concerning above mentioned methods possible modification should be stated. In considered cases of design based on post-buckling state skin thicknesses $\delta_{post\text{-}buck}$ were defined for ultimate level. Provided that SP buckling is unacceptable with limit loading linear stability tasks should be considered and thicknesses δ_{fat} should be calculated and further on maximum thickness value should be selected based on more strict design requirement.

5. Conclusion

Shown methods of sandwich panels design based on post-buckling state with compression and shear can be used for aircraft structures design with limitations in stability and load-bearing capacity. Given height and filler characteristics are taken into account in basic correlations.

References

[1] Volmir A 1956 *Flexible Plates And Shells* (Moscow: Gostekhizdat)
[2] Mitrofanov O 2003 *Engineer Composite Load-Bearing Panels Design Methods* (Moscow: Sputnik+)
[3] Endogur A, Vainberg M and Ierusalimskiy K 1989 *Honeycomb Structures. Parameters Selection and Design* (Moscow: Mashinostroenie)

IOP Conf. Series: Materials Science and Engineering **684** (2019) 012021 doi:10.1088/1757-899X/684/1/012021

On an optimization method of chemical reactors

I D Burlakov[1] and A S Buldaev[2]

Buryat State University named after D. Banzarov, 24a Smolin Street, Ulan-Ude 670000, Russia

e-mail: [1]ivan.burlakov.91@mail.ru, [2]buldaev@mail.ru

Abstract. The class of model problems of optimal process control of chemical reactors is considered. A new method of constructing successive approximations to the optimal problem solution is proposed, based on the implementation of the constructed conditions for improving control in the form of fixed-point problems. This form allows to apply and modify the theory and methods of searching for fixed points to optimize control. A comparative analysis of the effectiveness of the proposed optimization method with well-known computational algorithms in the calculation of specific model problems of optimization of chemical reactors is carried out.

1. Introduction

The theory and methods of chemical reactors optimization based on mathematical models are of considerable interest in the field of chemical materials science and have an extensive bibliography [1-3]. The optimization problems under consideration are characterized by various features, such as the structure of the system of equations and constraints, the dimension of the state space, types of nonlinearities, multicriteria etc. At present, there is no universal computational procedure that is sufficiently effective for calculating various optimization problems. Therefore, the development of effective optimization methods focused on taking into account the features of classes of applied problems is relevant.

The theory of optimal control was developed in various schools of thoughts, including the science of our country that made a significant contribution to its formation (A. A. Feldbaum, L. S. Pontryagin, A. A. Milyutin, and their followers). A number of numerical and approximate methods were developed for solving optimal control problems that continue to develop in different directions (R. Gabasov, F. M. Kirillova, V. F. Krotov, V. I. Gurman, F. L. Chernousko, R. P. Fedorenko [4], et al.).

In recent years, the works of V. A. Srochko [5] and A. S. Buldaev [6-8] have actively developed nonlocal methods for improving control as opposed to local methods (for example, gradient methods) do not use a laborious operation varying control at each iteration of methods in a sufficiently small neighborhood of control.

2. Optimization task and method

The class of optimization problems for chemical reactors, reduced to the following general form, is considered:

$$\dot{x}(t) = f\left(x(t), u(t), t\right), \quad x(t_0) = x_0, \tag{1}$$

$$u(t) \in U, \quad t \in T = [t_0, t_1], \tag{2}$$

$$\Phi(u)=\varphi(x(t_1))+\int_{t_0}^{t_1} F(x(t),u(t),t)dt \to \min_{u\in V}, \tag{3}$$

where $x(t)=(x_1(t),...,x_n(t))$ is the state vector of the system, $u(t)=(u_1(t),...,u_m(t))$ is the vector of controlled variables. As admissible controls, we consider the set V of piecewise continuous functions on $[t_0,t_1]$ with values in a convex closed set $U\subset R^m$. The time interval T and the initial state x_0 are set.

Using the Lagrange method or the penalty method, problems with terminal, phase, and mixed constraints can be reduced to the form (1)-(3).

In the class of problems under consideration, it is assumed that the function $\varphi(x)$ is continuously differentiable on the set R^n, the vector functions $F(x,u,t)$, $f(x,u,t)$ and their derivatives $F_x(x,u,t)$, $F_u(x,u,t)$, $f_x(x,u,t)$, $f_u(x,u,t)$ are continuous in the aggregate of arguments (x,u,t) on the set $R^n\times U\times T$.

Consider the Pontryagin function with an adjoint variable p

$$H(p,x,u,t)=\langle p,f(x,u,t)\rangle-F(x,u,t),$$

and in accordance with [7] introduce a modified differential-algebraic conjugate system:

$$\dot p(t)=-H_x(p(t),x(t),w(t),t)-r(t), \tag{4}$$

$$p(t_1)=-\varphi_x(x(t_1))-q, \tag{5}$$

$$H(p(t),y(t),w(t),t)-H(p(t),x(t),w(t),t)=\langle H_x(p(t),x(t),w(t),t)+r(t),y(t)-x(t)\rangle, \tag{6}$$

$$\varphi(y(t_1))-\varphi(x(t_1))=\langle\varphi_x(x(t_1))+q,y(t_1)-x(t_1)\rangle, \tag{7}$$

in which, by definition, $r(t)\equiv 0$, $q=0$ is assumed in the case of the $\varphi(x)$, $F(x,u,t)$, $f(x,u,t)$ functions linear in the variable x (state-linear problem (1)-(3)), as well as in the case of $x(t)=y(t)$ with the corresponding $t\in T$.

In the state-linear problem (1)-(3), the modified conjugate system (4)-(7) by definition coincides with the standard conjugate system.

In the state nonlinear problem (1)-(3), the algebraic equations (6) and (7) can always be analytically solved with respect to the values of $r(t)$ and q in the form of explicit or conditional formulas (possibly not the only way) [7].

Thus, the differential-algebraic conjugate system (4)-(7) can always be reduced (possibly not in a unique way) to a differential conjugate system with uniquely defined values of $r(t)$ and q.

To control $v\in V$, we denote $x(t,v)$, $t\in T$ is the solution of system (1), (2) for $u(t)=v(t)$. For a pair of controls $u\in V, v\in V$, we denote $p(t,u,v)$, $t\in T$ is the solution of the conjugate system (4)-(7) for $w(t)=u(t)$, $x(t)=x(t,u)$, $y(t)=x(t,v)$.

Consider the problem of improving control in the following form: for a given control $u\in V$, it is required to find a control $v\in V$ with the condition of $\Phi(v)\le\Phi(u)$.

Let us denote the P_U is the operator of projection onto the set U in the Euclidean norm:

$$P_U(z) = \arg\min_{w \in U}\left(\|w - z\|\right).$$

In accordance with [7], the projection conditions for improving the control $u \in V$ with the projection parameter $\alpha > 0$ in problem (1), (2) take the form:

$$v(t) = P_U\left(u(t) + \alpha\left(H_u\left(p(t,u,v), x(t,v), u(t), t\right) + s(t)\right)\right),\ t \in T. \tag{8}$$

$$H\left(p(t,u,v), x(t,v), v(t), t\right) - H\left(p(t,u,v), x(t,v), u(t), t\right) = \left\langle H_u\left(p(t,u,v), x(t,v), u(t), t\right) + s(t), v(t) - u(t)\right\rangle, \tag{9}$$

where in equation (9), by definition, $s(t) = 0$ is assumed in the case of the $F(x,u,t)$, $f(x,u,t)$ functions linear in the variable u (linear control problem (1)-(3)), as well as in the case of $u(t) = v(t)$ with the corresponding $t \in T$.

For the linear control problem (1)-(3), system (8), (9) by definition reduces to equation (8) with $s(t) = 0$, $t \in T$.

For a non-linear control problem, equation (9) can always be analytically solved respectively with the value of $s(t)$ (perhaps not the only way).

Thus, system (8), (9) can always be reduced to an equation of the form (8) with a uniquely determined quantity $s(t)$.

According to [7], the solution of system (8), (9) provides improved control $u \in V$ for any parameter $\alpha > 0$ with an estimate of the improvement:

$$\Phi(v) - \Phi(u) \le -\frac{1}{\alpha}\int_T \|v(t) - u(t)\|^2 \, dt.$$

Moreover, the improvement of control is guaranteed not only in a sufficiently small neighborhood of control $u \in V$. Thus the improvement procedure based on the solution of system (8), (9) has the property of nonlocality, in contrast to gradient and other local methods for improving control.

Conditions (8), (9) for a given method of unambiguous resolving equations (6), (7) and (9), respectively, quantities $r(t)$, q and $s(t)$ can be considered as a fixed point problem for the control operator uniquely determined by the right-hand side of equation (8).

To solve fixed point problems, one can use well-known simple iteration methods and their modifications [9]. As applied to the problem (8), (9), a simple iteration algorithm for $k \ge 0$ has the form:

$$v^{k+1}(t) = P_u\left(u(t) + \alpha\left(H_u\left(p\left(t,u,v^k\right), x\left(t,v^k\right), u(t), t\right) + s(t)\right)\right), \tag{10}$$

$$H\left(p\left(t,u,v^k\right), x\left(t,v^k\right), v^k(t), t\right) - H\left(p\left(t,u,v^k\right), x\left(t,v^k\right), u(t), t\right) = \left\langle H_u\left(p\left(t,u,v^k\right), x\left(t,v^k\right), u(t), t\right) + s(t), v^k(t) - u(t)\right\rangle,\ t \in T. \tag{11}$$

For $k = 0$ the initial approximation $v^0 \in V$ is specified for the iterative process, which, in the practice of calculations, is usually chosen to control $u \in V$.

Iterating for the index $k \geq 0$ is conducted until the first strictly improvement in control $u \in V$. Next, a new fixed-point problem is constructed to improve the obtained projection control and the iterative process is repeated.

If a strict control improvement does not occur, then the iterative process is carried out until the condition:

$$\left|\Phi\left(v^{k+1}\right)-\Phi\left(v^{k}\right)\right| \leq \left|\Phi\left(v^{k}\right)\right| \cdot \varepsilon,$$

where $\varepsilon > 0$ is the specified accuracy of the calculation. At this point, the construction and calculation of successive problems for improving control ends.

3. Examples

A comparative analysis of the effectiveness of the proposed optimization method is carried out on specific model problems of reactor optimization known in the literature. The computational implementation of the optimization method has the following features.

The numerical solution of phase and conjugate Cauchy problems is carried out by the fifth or sixth order accuracy Runge – Kutta – Werner method using IMSL library of the Fortran PowerStation 4.0 language [10]. Values of controlled, phase and conjugate variables are stored in nodes of a fixed uniform grid T_h on the interval T with a given sampling step $h > 0$. In the intervals between adjacent nodes of the grid T_h, the value of the control function was assumed to be constant and equal to the value in the left node.

3.1. Example 1

The problem of a chemical reactor optimization is considered [4,11]:

$$\begin{cases} \dot{x}_1(t) = -\left(k_1(u)+k_2(u)+k_3(u)\right)x_1, & x_1(0)=1, \\ \dot{x}_2(t) = k_1(u)x_1 - k_4(u)x_2, & x_2(0)=0, \\ \dot{x}_3(t) = k_4(u)x_2 - k_5(u)x_3, & x_3(0)=0. \end{cases}$$

$$u(t) \in U = [0, 823],\ t \in T = [0, 1];\ \Phi(u) = -x_3(1) \to \min.$$

The system of differential equations describes the reactions that occur in a mixture of three substances, $x_i(t), i = 1,2,3$ is their concentration. The first substance is raw materials, the second is an intermediate product, and the third is the final product. The temperature, on which the reaction intensities depend, is considered as a control $u(t)$.

The functions $k_i(u)$ have the characteristic form of chemical kinetics

$$k_i(u) = C_i \exp\left[\frac{E_i}{R}\left(\frac{1}{658} - \frac{1}{u}\right)\right],\quad i = 1,\ldots,5.$$

The values of the constants are taken from the book [12]: $C_1 = 1.02$, $C_2 = 0.93$, $C_3 = 0.386$, $C_4 = 3.28$, $C_5 = 0.084$, $R = 1.9865$, $E_1 = 16000$, $E_2 = 14000$, $E_3 = 15000$, $E_4 = 10000$, $E_5 = 15000$.

In [4], this problem was solved using the gradient projection method (MPG), in [11] it was done with the conditional gradient method (MCG) and the conditional quasigradient method of the first order (MCG-1).

To solve the problem with the proposed optimization method, we determine the necessary structures.

The Pontryagin function has the form:

$$H(p,x,u,t)=-p_1\left[k_1(u)+k_2(u)+k_3(u)\right]x_1+p_2\left[k_1(u)x_1-k_4(u)x_2\right]+p_3\left[k_4(u)x_2-k_5(u)x_3\right].$$

Since the problem is linear in the state, the differential-algebraic conjugate system coincides with the standard conjugate system and has the form:

$$\dot{p}_1(t)=p_1\left[k_1(u)+k_2(u)+k_3(u)\right]-p_2k_1(u),\ p_1(1)=0,\ \dot{p}_2(t)=p_2k_4(u)-p_3k_4(u),\ p_2(1)=0,$$

$$\dot{p}_3(t)=p_3k_5(u),\ p_3(1)=1.$$

To solve the fixed point problem, having the form (8), (9), an iterative process in the form of (10), (11) was used. In this case, the value $s(t)$ is determined by the following rule:

$$s(t)=\begin{cases}\dfrac{H\left(p^k(t),x^k(t),v^k(t),t\right)-H\left(p^k(t),x^k(t),u(t),t\right)}{v^k(t)-u(t)}-H_{\mathrm{u}}\left(p^k(t),x^k(t),u(t),t\right),\ v^k(t)\neq u(t),\\ 0,\ v^k(t)=u(t).\end{cases}$$

Here we introduce the notation $x^k(t)=x\left(t,v^k\right)$, $p^k(t)=p\left(t,u,v^k\right)$.

In the initial problem of improving control, $u(t)\equiv 600$ was chosen as the initial (starting) control.

Table 1 shows comparative results with the known methods of calculating the problem by the proposed fixed-point method (FPM) for various values of the parameter α. Φ^* is the estimated value of the functional. N is the total number of calculated phase and conjugate Cauchy problems.

Table 1.

Method	Φ^*	N
MCG	-0.43620	876
MCG–1	-0.43682	215
MPG	-0.435	–
FPM $\left(\alpha=10^4\right)$	-0.43735	47
FPM $\left(\alpha=10^3\right)$	-0.43558	155

As the parameter decreases from $\alpha=10^3$, the convergence of the iterative process slows down and the total number of Cauchy computational problems increases. As the parameter increases to $\alpha=10^5$, the iterative process of the FPM ceases to converge.

Figures 1 and 2 illustrate the calculated control and the corresponding phase trajectories obtained by the fixed-point method for the parameter $\alpha=10^4$.

3.2. Example 2

The stabilization of a chemical reactor is considered; it is an apparatus with a stirrer and a supply channel for refrigerant [13]:

$$\Phi(u)=\int_0^{0.78}\left(x_1^2(t)+x_2^2(t)\right)dt\to\min,\ \dot{x}_1(t)=-2(x_1+0.25)+(x_2+0.5)e^{\frac{25x_1}{x_1+2}}-(x_1+0.25)u,$$

$$x_1(0)=0.05,\ \dot{x}_2(t)=0.5-x_2-(x_2+0.5)e^{\frac{25x_1}{x_1+2}},\ x_2(0)=0,\ x_1(0.78)=0,$$

$$x_2(0.78)=0,\ u(t)\in[-1,1],\ t\in[0,0.78].$$

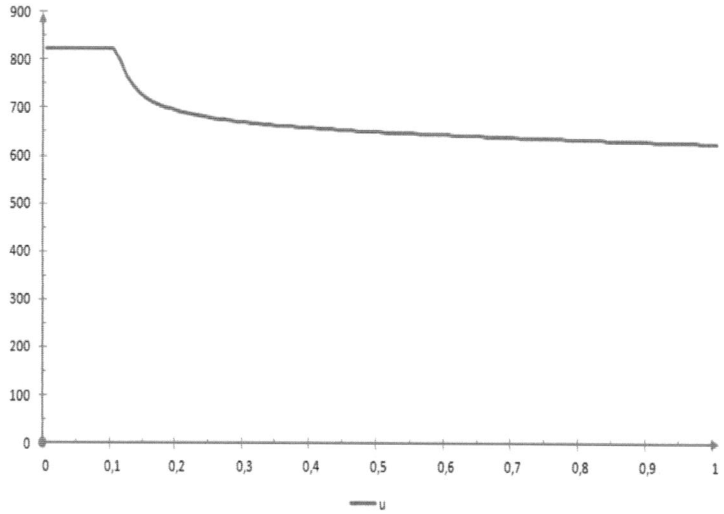

Figure 1. Control.

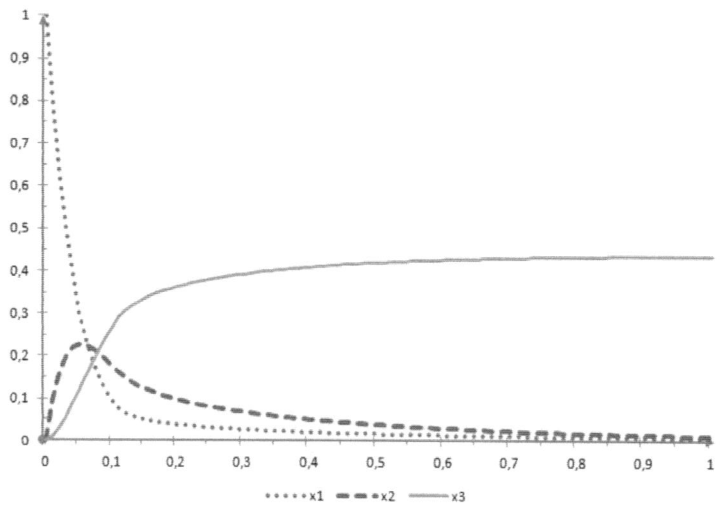

Figure 2. Phase trajectories.

The functions $x_1(t)$ and $x_2(t)$ respectively the deviations of the temperature and concentration of the reacting mixture. The control $u(t)$ characterizes the flow of refrigerant that regulates the irreversible exothermic reaction.

In [13], the problem reduces to the finite-dimensional optimization problem using discretization by state variables and control and replacing derivatives with finite differences according to the Euler scheme.

We apply the proposed method to solve the continuous problem of optimal control under consideration. We reduce the problem to a problem of the form (1)-(3) without terminal restrictions by introducing the penalty functional $\Phi^M(u)$ with a sufficiently large penalty parameter $M>0$:

$$\Phi^M(u)=x_3(0.78)+M\left(x_1^2(0.78)+x_2^2(0.78)\right)\to\min,$$

$$\dot{x}_1(t)=-2(x_1+0.25)+(x_2+0.5)e^{\frac{25x_1}{x_1+2}}-(x_1+0.25)u,\ x_1(0)=0.05,$$

$$\dot{x}_2(t)=0.5-x_2-(x_2+0.5)e^{\frac{25x_1}{x_1+2}},\ x_2(0)=0,\ \dot{x}_3(t)=x_1^2+x_2^2,\ x_3(0)=0,$$

$$u(t)\in[-1,1],\ t\in[0,0.78].$$

The Pontryagin function for the penalty problem has the form:

$$H(p,x,u,t)=p_1\left[-2(x_1+0.25)+(x_2+0.5)e^{\frac{25x_1}{x_1+2}}-(x_1+0.25)u\right]+p_2\left[0.5-x_2-(x_2+0.5)e^{\frac{25x_1}{x_1+2}}\right]$$

$$+p_3\left[x_1^2+x_2^2\right].$$

The modified conjugated system (4)-(7) by the rules of the unambiguous choice of the quantities $r(t)$ and q is reduced to the system:

$$\dot{p}_1=-H_{x_1}-r_1(t),\ p_1(0.78)=-Mx_1(0.78)-My_1(0.78),\ \dot{p}_2=-H_{x_2}-r_2(t),$$

$$p_2(0.78)=-Mx_2(0.78)-My_2(0.78),\ p_3\equiv-1,$$

wherein

- if $y_1(t)\neq x_1(t)$, then $r_1(t)=\dfrac{\Delta_y H}{y_1(t)-x_1(t)}-H_{x_1}-H_{x_2}\dfrac{y_2(t)-x_2(t)}{y_1(t)-x_1(t)}$, $r_2(t)=0$;
- if $y_1(t)=x_1(t)$, then
- a) if $y_2(t)\neq x_2(t)$, then $r_1(t)=0$, $r_2(t)=\dfrac{\Delta_y H}{y_2(t)-x_2(t)}-H_{x_2}$;
- b) if $y_2(t)=x_2(t)$, then $r_1(t)=0$, $r_2(t)=0$.

The notation $\Delta_y H=H(p,y,w,t)-H(p,x,w,t)$ is introduced here.

Due to the linearity of the optimal control problem for control, system (8), (9) reduces to the fixed point problem (8) with $s(t)\equiv 0$, to solve which the simple iteration algorithm was used for $k\geq 0$:

$$v^{k+1}(t)=P_U\left(u(t)+\alpha H_u\left(p\left(t,u,v^k\right),x\left(t,v^k\right),u(t),t\right)\right),\ t\in T.$$

In the initial problem of improving control, $u(t)\equiv 1$ was chosen as the starting control.

The calculation by the proposed method of fixed points (FPM) was carried out for various values of the parameter $\alpha>0$ and the penalty coefficient $M>0$. The convergence of the FPM for various values of $M>0$ depended on the choice of the parameter $\alpha>0$. For example, for $M=2$, the FPM converged at $\alpha < 0.4$. Moreover, for small $\alpha > 0$ the convergence rate was slow. The best speed, estimated by the number of solved Cauchy problems, was observed for $\alpha \in (0.10; 0.14)$. With an increase in the penalty parameter for convergence, we had to decrease α. Moreover, the convergence rate also slowed down. The best-calculated value of the target functional Φ^* was achieved for the parameters $\alpha = 0.13,\ M = 2$.

Table 2 provides a comparative analysis of the calculation results by the proposed FPM method and the gradient projection method (MPG) used in [13] to calculate a finite-dimensional analog of the problem.

Table 2.

Method	Φ^*	$x_1^*(0.78)$	$x_2^*(0.78)$	N
FPM	0.00200	$-9.62 \cdot 10^{-4}$	$1.24 \cdot 10^{-3}$	90
MPG	0.00220	$-6.16 \cdot 10^{-6}$	$-0.63 \cdot 10^{-6}$	–

Figures 3 and 4 show the calculated control $u(t)$ and the corresponding phase trajectories $x_1(t)$, $x_2(t)$, obtained by the FPM method.

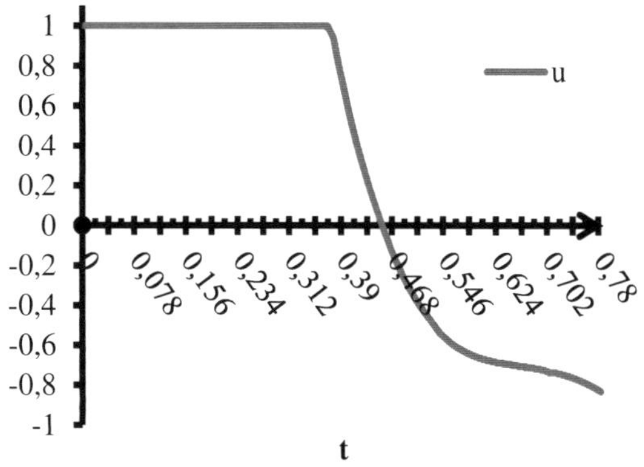

Figure 3. Control.

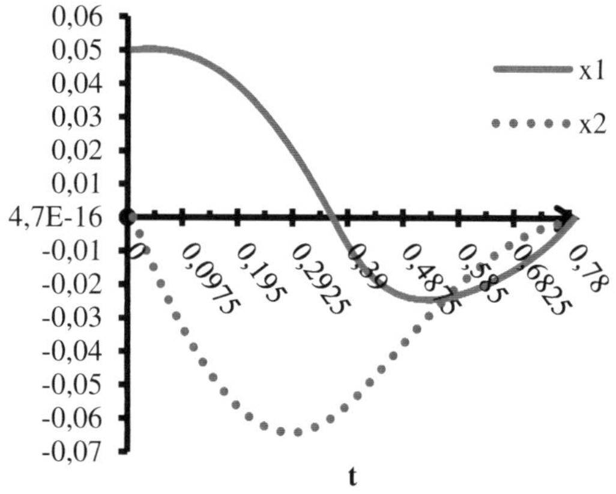

Figure 4. Phase trajectories.

4. Conclusion

The conducted comparative computational experiments on model optimization problems of chemical reactors demonstrate the computational and qualitative efficiency of the proposed fixed point method acceptable for practice in comparison with known methods. The developed method for finding approximately optimal solutions has a fairly wide convergence region in the initial approximation and is characterized by the convenience and simplicity of the experimental setup of a scalar projection parameter that regulates the quality and convergence rate of the iterative process under consideration. In general, the optimization method based on the calculation of the constructed fixed-point problems is reduced to alternating solutions of Cauchy problems for phase and conjugate variables of the problem.

The indicated properties of the proposed fixed-point method are important factors in improving the computational and qualitative efficiency of solving optimization problems of chemical reactors.

Acknowledgments
This work was supported by the Ministry of Education and Science of the Russian Federation, project 1.5049.2017BC, and Russian Foundation for Basic Research, project 18-41-030005-r-a.

IOP Conf. Series: Materials Science and Engineering **684** (2019) 012021 doi:10.1088/1757-899X/684/1/012021

References

[1] Boyarintsev A I and Kafarov V V 1969 *Optimization Methods in Chemical Technology* (Moscow: Chemistry) p 563
[2] Ostrovsky G M and Berezhinsky T A 1984 *Optimization of Chemical-Technological Processes. Theory and Practice* (Moscow: Chemistry) p 240
[3] Bochkarev V V 2014 *Optimization of Chemical-Technological Processes* (Tomsk: Tomsk Polytechnic University Press) p 264
[4] Fedorenko R P 1978 *Approximate Solution of Optimal Control Problems* (Moscow: Nauka) p 487
[5] Srochko V A 2000 *Iterative Methods for Solving Optimal Control Problems* (Moscow: Nauka) p 160
[6] Buldaev A S 2008 *Perturbation Methods in Problem of the Improvement and Optimization of the Controlled Systems* (Ulan-Ude: Buryat State University Press) p 260
[7] Buldaev A S 2017 Fixed-point methods based on the projection operations in problems of optimization of control functions and parameters of dynamical systems *Buryat State University Bulletin. Mathematics, Informatics* **1** 38–54
[8] Buldaev A S and Burlakov I D 2018 About one approach to the numerical solution of nonlinear optimal speed problems *Bulletin of the South Ural State University. Series "Mathematical Modelling, Programming and Computer Software"* **11(4)** 55–66
[9] Samarsky A A and Gulin A V 1989 *Numerical Methods* (Moscow: Nauka) p 432
[10] Bartenev O V 2001 *Fortran for Professionals. Mathematical Library IMSL. Part 2* (Moscow: Dialog-MIFI) p 320
[11] Srochko V A, Antonik V G and Mamonova N V 2007 Computational comparison of gradient type methods in optimal control problems *Bulletin of Irkutsk State University. Series "Mathematics"* **1** 275–90
[12] Rosenbrock H H and Storey C 1966 *Computational Techniques for Chemical Engineers* (Oxford: Pergamon Press) p 346
[13] Kirk D E 2004 *Optimal Control Theory. An Introduction* (New York: Dover Publications) p 452

Determination of the geometric characteristics of fatigue cracks by digital image processing method

S A Bazaron[1], T Ts Damdinova[2] and E B Bochektueva

East Siberia State University of Technology and Management, 40v Klyuchevskaya Street, Ulan-Ude 670013, Russia

e-mail: [1]seseg003@gmail.com, [2]damdinova.tatyanac@gmail.com

Abstract. The article deals with the problem of determining the rate of crack growth in the process of fatigue failure of helicopter part samples. The studies were conducted using a video recording of the experiment. Methods of digital image processing were applied to the frames of the video recording to process the data of the object destruction process. Based on the developed algorithm, computational experiments were carried out, the geometric parameters of the cracks were determined for calculating the rate of change in their sizes. The use of digital image processing methods allows significantly improve the accuracy and reduce the processing time of experimental data.

1. Introduction

In all branches of engineering, especially where accidents lead to catastrophic consequences due to the destruction of critical parts, there is a problem of timely detection and prevention of fatigue failure. Mechanisms of wear and failure of materials in various industries are well studied and presented in many scientific articles [1], there is also a large number of publications on this problem for various materials with modeling of the fracture process [2-4].

Continuous progress in the field of computer science and applied mathematics has allowed modeling to take a leading position in the design and production of modern technology. Modeling in the production of complex equipment allows to avoid a fatal error in the safety of the aircraft, or it is used after production as a maintenance tool to identify fatigue, vulnerability and degree of equipment parts destruction. To determine the main causes that affect the destruction of materials and improve the strength characteristics, it is necessary to conduct numerous experiments. Specialized programs are used for experimental data processing, their testing is carried out and optimal parameters and characteristics satisfying the specified conditions are determined [5-8].

To detect fatigue of the objects under study nondestructive testing methods are widely used [9]. The development of modern means of obtaining images, reducing their distortion, well-developed algorithms of digital image processing methods are actively used to solve this problem [10-13].

2. Objects and research methods

The paper [14] describes a technique for determining the characteristics of cracks in aircraft products. Based on this work, a method for processing the video of the crack development in the helicopter parts was developed and implemented. The object of the study is the spar of the rotor blade of the Mi-171 helicopter. The component of the helicopter was under axial tensile stress of 58.86 MPa and a bending

IOP Conf. Series: Materials Science and Engineering **684** (2019) 012022 doi:10.1088/1757-899X/684/1/012022

moment stress of 34.34 MPa. The experiment was videotaped with a GoPro HERO 4 Silver Edition action camera. The duration of the experiment is 185 hours.

At the stage of pre-processing of data from the video of crack development, the frames were selected in order to highlight cracks in the images by digital image processing methods. Image analysis was performed according to the following algorithm:

- binarization, i.e. conversion of the original image to black and white for object and background;
- determination of the crack boundaries;
- determination of the geometric parameters of the crack (length, area);
- calculation of crack growth rate.

3. Image processing results

To study crack growth rate, it is necessary to obtain information on size of a crack. Fast and accurate source image data processing plays a key role here.

The grayscale image of the crack is shown in figure 1.

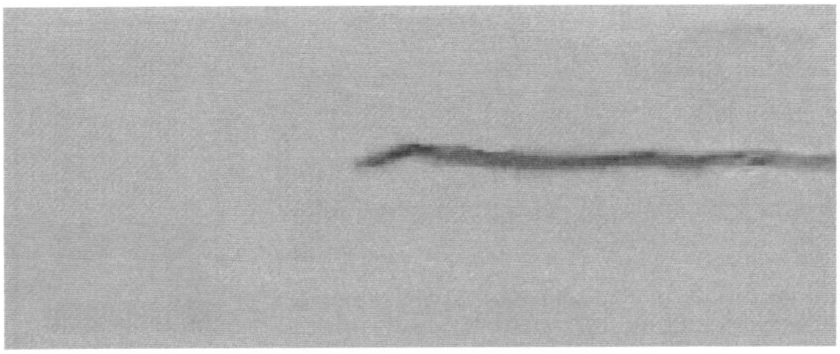

Figure 1. F – the grayscale image of the crack.

For binarization and determination of crack boundaries, a computer program has been designed using Python programming language. The binarization threshold which is used in (1) was taken as $T = 110$ (figure 2).

$$B = \begin{cases} 1 & \text{if } f[i,j] \geq T; \\ 0 & \text{otherwise,} \end{cases} \tag{1}$$

where $f[i,j] \in F$ is the matrix element of the source grayscale image.

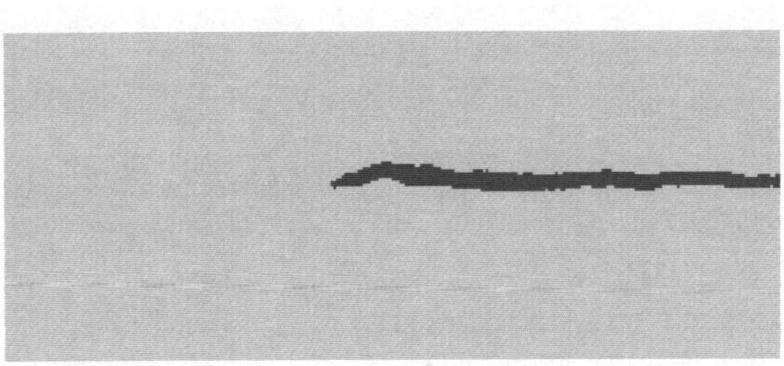

Figure 2. B – the binary image of the crack.

To determine the boundary pixels of the crack, vertical and horizontal Sobel gradient filters have been implemented [10, 11]. As a result of this stage of the algorithm, an image matrix V with boundary pixels was obtained for each image.

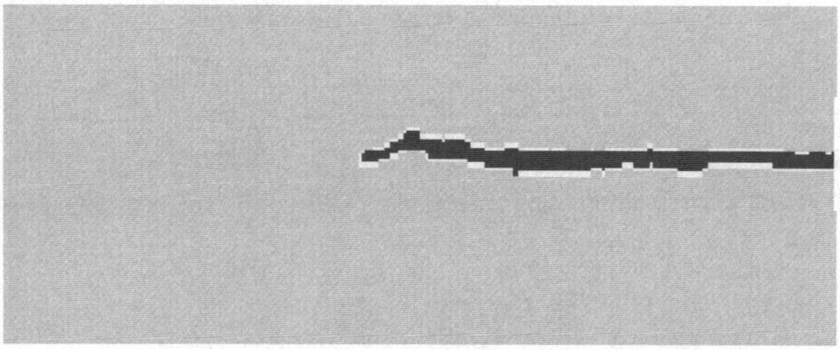

Figure 3. V – the image of the boundary pixels.

Quantity of pixels of perimeter P is calculated according to the formula (2), and value S as quantity of the crack area is determined by the formula (3).

$$P = \begin{cases} \sum_i \sum_j 1 & \text{if } v[i,j] \neq 0; \\ 0 & \text{otherwise,} \end{cases} \tag{2}$$

where $v[i,j] \in V$ is the matrix element of the boundary image.

$$S = \begin{cases} \sum_i \sum_j b[i,j] & \text{if } b[i,j]=1; \\ 0 & \text{otherwise,} \end{cases} \tag{3}$$

where $b[i,j] \in B$ is the element of the binary matrix.

The coordinates of the crack center of gravity were calculated by the formulas bellow and the center point is shown in figure 4 as a red spot with axes of inertia:

$$x_c = \frac{\sum_i \sum_j j \text{ if } b[i,j]=1}{S}; \; y_c = \frac{\sum_i \sum_j i \text{ if } b[i,j]=1}{S}.$$

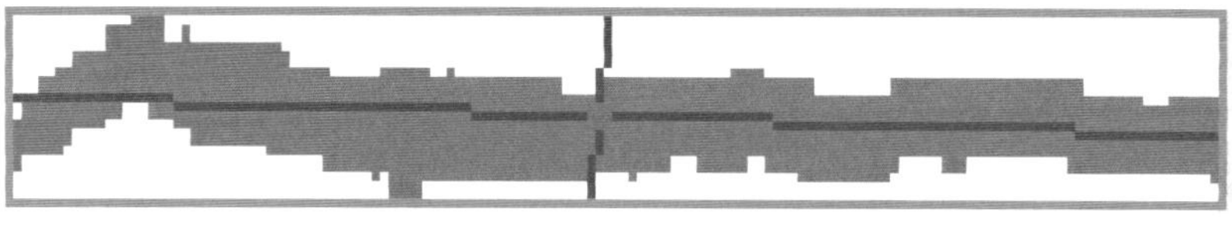

Figure 4. The crack centre of gravity and inertia axes.

Using the coordinates of the center of gravity, the angle of inclination of the axis of inertia is calculated. The calculation results of the program for determining the crack characteristics are presented below in table 1.

Table 1. Crack geometry.

Area *S* (px)	Perimeter *P* (px)	Center of gravity coordinates	The inertia axis angle
417	197	$x = 240$; $y = 61$	–0.827

The developed program determines geometric characteristics from the binary image of the crack – area, perimeter, coordinates of the center of gravity, and the angle of inclination of the axis of inertia. The calculated characteristics enable automatically monitor the behavior and the growth rate of cracks. Compared to the manual crack measurement method, the time and effort required for fatigue testing can be significantly reduced. The use of digital image processing methods improves the accuracy of calculations and effectively manage the safety and residual life of the objects under study.

Acknowledgments
The work was supported by State task of the Ministry of Science and Higher Education of Russian Federation, project No. 9.7667.2017/8.9, and by the grant "Young Scientists of ESSUTM – 2019".

References
[1] Dasgupta A and Pecht M 1991 Material failure mechanisms and damage models *IEEE Transactions on Reliability* **40(5)** 531–6
[2] Volegov P S, Gribov D Sand Trusov P V 2015 Damage and fracture: review of experimental studies *Fizicheskaya mezomekhanika* **3** 11–24
[3] Burov A E 2008 Models of destruction of fibrous composite materials *Bulletin of the Siberian State Aerospace University after M.A. Reshetnev* **3(20)** 133–8
[4] Habtour E, Drake G S, Dasgupta A, Al-Bassyiouni M and Choi C 2010 Improved reliability test using multiaxial electrodynamics vibration *IEEE Xplore Digital Library* 5448050
[5] Samusenko A G, Esaulov S Yu and Ivchin V A 2012 Simulation of helicopter vibration reduction by means of higher harmonics control with the use of flight simulator *Proc. of 38th European Rotorcraft Forum (Amsterdam)* (Amsterdam: Netherlands Association of Aeronautical Engineers Press) pp 104–22
[6] Markov A, Barabanov A and Tsirlov V 2018 Models for testing modifiable systems *Probabilistic Modeling in System Engineering* ed A Kostogryzov (London: IntechOpen) chapter 7 pp 147–68
[7] Cheung C, Valdés J J and Puthuparampil J 2016 Improving load signal and fatigue life estimation for helicopter components using computational intelligence techniques *Recent Advances in Computational Intelligence in Defense and Security* (*Studies in Computational Intelligence* vol 621) ed R Abielmona, R Falcon et al. (Cham: Springer) pp 605–34
[8] Barabanov A V, Marcov A S and Tsirlov V L 2018 Statistics of software vulnerabiliry detection in certification testing *Journal of Physics: Conference Series* **1015** 042033
[9] Oster R 2012 Non-destructive testing methodologies on helicopter fiber composite components challenges today and in the future *18th World Conf. on Non-Destructive Testing (Durban)* (Northampton: British Institute of Non-Destructive Testing Press) pp 2361–70
[10] Soifer V A et al. 2006 *Introduction to Digital Processing of Signals and Images: Improving the Quality and Evaluation of Geometric Parameters of Images* (Samara: Samara State Aerospace University Press)
[11] Gonzalez R C and Woods R E 2007 *Digital Image Processing* (Upper Saddle River, New Jersey: Pearson Prentice Hall)
[12] Bernd J and Horst H 2000 *Computer Vision and Applications. A Guide for Students and Practitioners* (San Diego: Academic Press)
[13] Damdinova T 2014 Methods of digital images creation and analysis of their aberrations *Vopr. Kiberbezop.* **5(8)** 43–6
[14] Bokhoeva L A, Kurokhtin V Yu and Rogov V E 2016 Investigation of cracks growth in

aerotechnics products on the basis of full-scale tests *BSU bulletin. Chemistry. Physics* **2-3** 63–8

IOP Conf. Series: Materials Science and Engineering **684** (2019) 012023 doi:10.1088/1757-899X/684/1/012023

Experimental study of continuous welded rails deformation under the effect of axial force

P V Novoseltsev and L A Bokhoeva

East Siberia State University of Technology and Management, 40v Klyuchevskaya Street, Ulan-Ude 670013, Russia

e-mail: nov-pv@mail.ru

Abstract. On the Russian railways there is a transition from conventional rail to a continuous welded rail track, however during its operation, occasional track disturbances occur due to internal stresses in the rails caused by the temperature and longitudinal stresses generated by passing rolling stock and creepage of the railway. Stresses can be tensile in winter and compressive in summer. They can reach dangerous values leading to rupture of the track at low temperatures and its ejection at high temperatures. To ensure traffic safety, it is necessary to develop a technology for monitoring stresses arising during operation. The authors created and obtained a patent for a device which made it possible to determine the generated stresses. Studies were carried out on welded rail behaviour and the dependence of displacements and stresses on the action of an external force was obtained.

1. Introduction

The author created "a device for determining the longitudinal stiffness of a rail track" (patent for utility model RU 138300 U1) [1]. With this device a longitudinal force P is applied to the rails, the displacement of the rails δ is determined and the stresses generated in the rails are calculated.

2. Methods

We study the option when compressive stresses (σ_t) appeared in rails due to low temperatures. The design scheme is presented in figure 1(a). The longitudinal force P is generated in the rail by the train, as a result, the distributed resistance force q acts on the rail from the embankment and the sleeper base (figure 1(b)) [2, 3].

The stresses generated in the rails are as follows:

- temperature stress σ_t (figure 1(c)) [2, 3];
- longitudinal stress on the force P (figure 1(d)) [4, 5];
- longitudinal stress caused by the distributed resistance force from the action of the upper structure of the track q (figure 1(e)) [2, 6].

It was accepted that:

- the force P acts at a distance of $2d$;
- the force from the action of the upper structure of the track on the rail has a linear relationship (figure 1(b)):

IOP Conf. Series: Materials Science and Engineering **684** (2019) 012023 doi:10.1088/1757-899X/684/1/012023

$$q_z = q_0 \cdot \frac{z}{d};$$

- compressive temperature stresses σ_t are evenly distributed in length (figure 1(c));

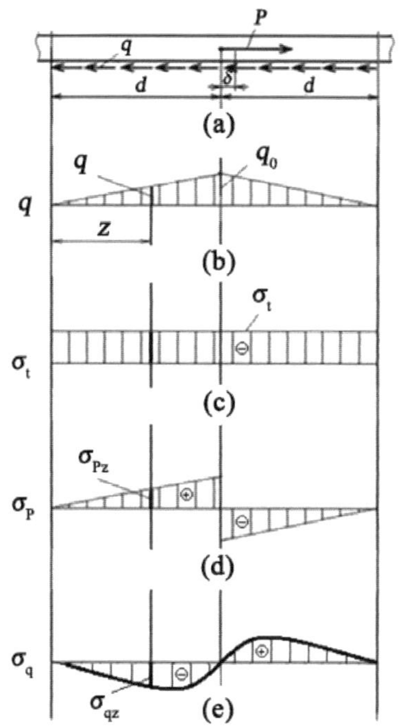

Figure 1.

- longitudinal stresses in the rail from the action of the force P are distributed according to a linear relationship (figure 1(d)):

$$p_z = \frac{P \cdot z}{2 \cdot d \cdot F};$$

- stress σ_q from the action of the railway track on the rail is determined by the dependence (figure 1(e)):

$$\sigma_{qz} = q_0 \cdot \frac{z^2}{d \cdot 2 \cdot F}.$$

The potential energy of deformation caused by the temperature is determined by the dependence:

$$U_0 = 2 \cdot \int_0^d \frac{\sigma_t^2 \cdot F^2 \, dz}{2 \cdot E \cdot F}. \quad (1)$$

The potential energy of deformation under the temperature σ_t, longitudinal force P, resistance force q:

$$U_1 = \int_0^d \frac{\left(-\sigma_t + \frac{P \cdot z}{2 \cdot d \cdot F} - q_0 \cdot \frac{z^2}{d \cdot 2 \cdot F}\right)^2 \cdot F^2 \, dz}{2 \cdot E \cdot F} + \int_0^d \frac{\left(-\sigma_t - \frac{P \cdot z}{2 \cdot d \cdot F} + q_0 \cdot \frac{z^2}{d \cdot 2 \cdot F}\right)^2 \cdot F^2 \, dz}{2 \cdot E \cdot F}. \quad (2)$$

The increment of potential strain energy:

$$\Delta U = U_1 - U_0. \quad (3)$$

After the conversion of equations (1), (2), (3) we obtain:

$$\Delta U = \frac{\sigma_t \cdot q_0 \cdot d^2}{3 \cdot E} + \frac{P^2 \cdot d}{12 \cdot E \cdot F} - \frac{q_0{}^2 \cdot d^3}{20 \cdot E \cdot F}. \quad (4)$$

The increment of the potential energy of deformation is equal to the sum of the work of external forces:

$$\Delta U = A_i^e. \quad (5)$$

Experimental studies were conducted on the dependence of the displacement δ of the point of application of force P on its magnitude (see the graph in figure 2).

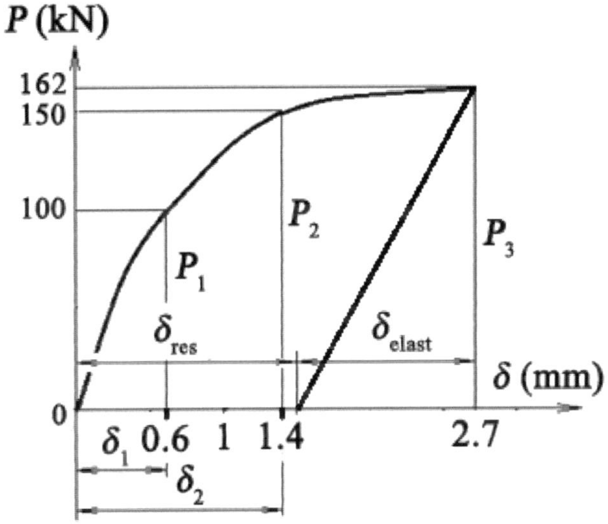

Figure 2. Experimental dependence of the movement of the rail track δ from the force P.

We represent the loading curve as a function:

$$P = a \cdot \delta^m. \quad (6)$$

We determine the coefficient m and, substituting P_1 and P_2, δ_1 and δ_2, we obtain:

$$m = \frac{\ln P_2 - \ln P_1}{\ln \frac{\delta_2}{\delta_1}}. \quad (7)$$

We express the coefficient a as:

$$a = \frac{P_1}{\delta_1^m}. \tag{8}$$

3. Results

In this section we calculate rail track parameters.

We find the sum of the work of external forces from formulas (6), (7), (8) and, having transformed, we get:

$$\sum A_i^{\mathrm{e}} = \int_0^{\delta} P(\delta) = \int_0^{\delta} a \cdot \delta^m = \frac{P \cdot \delta}{m+1}. \tag{9}$$

The sum of the forces is expended on the deformation of the rail track and the loss of energy during its movement:

$$\sum A_i^{\mathrm{e}} = A + A_{\mathrm{f}}.$$

Work spent on elastic deformation is as:

$$A = \frac{1}{2} \cdot \delta \cdot P,$$

where the displacement δ is found according to the graph in figure 2.

Work from energy loss:

$$A_{\mathrm{f}} = \sum A_i^{\mathrm{e}} - A = \frac{P \cdot \delta}{m+1} - \frac{1}{2} \cdot \delta \cdot P. \tag{10}$$

Elastic deformation is found by Hooke's law:

$$\delta = \frac{P \cdot d}{E \cdot F}.$$

Then, the length of the railway track d, deforming from the action of the force P:

$$d = \frac{\delta_{\mathrm{elast}} \cdot E \cdot F}{P}. \tag{11}$$

The work of strain energy loss is caused by the action of the resistance force of the upper structure of the path q:

$$A_{\mathrm{f}} = A_{\mathrm{q}} = 2\int_0^{d} q_{\mathrm{z}} \cdot \delta \cdot \frac{z}{d}\,\mathrm{d}z.$$

By transforming (11), and also expressing the length d (10), we obtain:

$$A_{\mathrm{f}} = A_{\mathrm{q}} = 2 \cdot q_0 \cdot \delta \cdot \frac{\delta \cdot E \cdot F}{P}.$$

By substituting the work of energy loss (10), we obtain the maximum value of the resistance force from the action of the upper structure of the path q_0:

$$q_0 = \frac{P^2 \cdot \left(\frac{\delta}{m+1} - \frac{1}{2} \cdot \delta_{\mathrm{elast}} \right)}{2 \cdot \delta \cdot \delta_{\mathrm{elast}} \cdot E \cdot F}.$$

After the transformation (5), (4) and (9), we obtain:

$$\frac{\sigma_t \cdot q_0 \cdot d^2}{3 \cdot E} + \frac{P^2 \cdot d}{12 \cdot E \cdot F} + \frac{q_0^2 \cdot d^3}{20 \cdot E \cdot F} = \frac{P \cdot \delta}{m+1}.$$

Defining the emerging stresses of the jointless track should be calculated as follows:

$$\sigma_t = \left(\frac{P \cdot \delta}{m+1} - \frac{P^2 \cdot d}{12 \cdot E \cdot F} - \frac{q_0^2 \cdot d^3}{20 \cdot E \cdot F} \right) \cdot \frac{3 \cdot E}{q_0 \cdot d^2}.$$

4. Conclusion

Using this method, you can:

- find emerging stresses in a rail in a dangerous place and compare it with permissible ones;
- find the shear resistance of the rail track and determine its technical condition.

Acknowledgments
The work was supported by State task of the Ministry of Science and Higher Education of Russian Federation, projects No. 9.7667.2017/8.9 and No. 9.11221.2018/11.12.

References

[1] Eliseev S, Novoseltsev V, Novoseltsev P, Suslov O and Gordeeva A 2014 *Device for Determining the Longitudinal Stiffness of the Rail Track: Russia patent No. 138300* (Moscow: Russian Federal Service for Intellectual Property)
[2] Yakovleva T, Karpushenko N, Klinov S, Putrya N and Smirnov M 1999 *Zheleznodorozhnyi Put* (Moscow: Transport)
[3] Novakovich V 2005 *Besstykovoi Put so Sverkhdlinnymi Relsovymi Pletyami* (Moscow: Marshrut)
[4] Novoseltsev V, Novoseltsev P and Gordeeva A 2013 *World of Transport and Transportation* **4(48)** 34–8
[5] Novoseltsev V, Dambaev Zh and Novoseltsev P 2007 *Modern Technologies. System Analysis. Modeling* **1(13)** 74–8
[6] Paramonov G P, Kovalevskii V N and Mysin A V 2018 Prognozirovanie razvala vzorvannoi gornoi massy na granitnom mestorozhdenii *International Journal of Mechanical Engineering and Technology* **9(11)** 1926–35

IOP Conf. Series: Materials Science and Engineering **684** (2019) 012024 doi:10.1088/1757-899X/684/1/012024

Accounting for shear deformations with bending a round plate

L A Bokhoeva[1] and A S Chermoshentseva[2]

[1] East Siberia State University of Technology and Management, 40v Klyuchevskaya Street, Ulan-Ude 670013, Russia
[2] Bauman Moscow State Technical University, 5/1 2nd Baumanskaya Street, Moscow 105005, Russia

e-mail: bohoeva@yandex.ru

Abstract. This paper presents a solution of the stability problem of a round plate taking into account the influence of shear deformations. The critical load of a round plate is determined taking into account shear deformations and without taking into account the shear deformations. The critical load taking into account the transverse shear with a local stability loss of a round plate is less than the critical load without taking into account the transverse shear. The inclusion of shear deformations in the study of the supercritical behavior of a round plate for isotropic materials does not have practical value, but it makes sense with extremely strong anisotropy of the elastic properties of the material.

1. Introduction

In the classical theory of plates, the assumption is used about a material element that is normal to the median surface before deformation [1]. This material element remains rectilinear and perpendicular to it even after deformation.

We used one of the simplest versions of an improved linear theory. The improved linear theory suggests that the material element of the plate is perpendicular to the median plane prior to deformation. This material element remains straight after bending the plate, but the rotation angles of this element are not directly related to the rotation angles of the normal to the median surface [2].

2. The mathematical model

We consider a plate with radius R and thickness h. The center of the plate coincides with the origin oxy. A thin axisymmetric plate is clamped along the contour. The plate is subject to a uniformly distributed load with intensity q. The deformation geometry of a round plate is described by deflection $w = w(r)$ and angle of rotation $\vartheta_r = \vartheta_r(r)$.

The deformation components of a layer located at a distance z from the median surface are

$$\varepsilon_r = -z\vartheta_r'; \varepsilon_\theta - z\vartheta_r/r; \varepsilon_z = 0;$$
$$\gamma_{xy} = 0; \gamma_{rz} = \vartheta_r - w'; \gamma_{\theta z} = 0,$$

where w' is the angle of rotation of the normal to the middle surface of the plate. Then the bending strain energy has the form

$$V=\pi\int\limits_0^R D\left[\left(\vartheta_r'+\vartheta_2/r\right)^2-2(1-\mu)\vartheta_r\vartheta_r'/r\right]\cdot r\,dr+\pi\int\limits_0^R B\left(\vartheta_\gamma-w'\right)^2 r\,dr,$$

where D is flexural stiffness of delamination, B is transverse shear stiffness:

$$D=\frac{E_1h^3}{12(1-\mu_{12}\mu_{21})},\quad B=G_{12}h\,.$$

We assume that the median surface of a round plate is inextensible. The change in the total potential energy of a round plate is determined by the expression

$$\Delta E=V-2\pi\int\limits_0^R q_X\frac{1}{2}\left(\frac{dw}{dr}\right)^2 r\,dr\,.$$

When constructing an approximate solution, we take

$$\vartheta_r=\Psi(r);\ w'=c\Psi(r),$$

where $\Psi(r)$ is a function that satisfies the boundary conditions of the task; C is parameter characterizing the magnitude of the transverse shear.

From the condition $\Delta E=0$ follows

$$q=\frac{\pi\int\limits_0^R D\left[\left(\Psi'+\Psi'/r\right)^2-2(1-\mu)\Psi'\Psi/r\right]\cdot r\,dr+(c-1)^2\int\limits_0^R B\Psi^2 r\,dr}{c^2\int\limits_0^R\Psi^2 r\,dr}. \tag{1}$$

We take $\Psi=\vartheta_r^0$, where ϑ_r^0 is the function corresponding to the solution of the problem without taking into account the transverse shift, then expression (1) has the form

$$q=\frac{q_{cr}^0+B(c-1)^2}{c^2}, \tag{2}$$

where q_{cr}^0 is the critical value of the load found without transverse shear, i.e.

$$q_{cr}^0=\frac{D_0\int\limits_0^R\left\{\left[\left(\vartheta_r^0\right)'+\vartheta_r^0/r\right]^2-2(1-\mu_{12})\cdot\left(\vartheta_r^0\right)'\cdot\vartheta_r^0/r\right\}r\,dr}{\int\limits_0^R\left(\vartheta_r^0\right)^2 r\,dr},\quad q_{cr}^0=16\frac{D}{R^2}.$$

Minimizing expression (2) with respect to the parameter C, we find

$$C_{cr}=1+q_{cr}^0/B,\ q_{cr}=\frac{q_{cr}^0}{1+q_{cr}^0/B}. \tag{3}$$

The critical load taking into account the transverse shear with a local loss of stability is slightly less than the Euler critical load [3]. Taking into account the effects of transverse shear leads to a decrease in the critical load. In particular, for an isotropic material at $G=\dfrac{E}{2(1+\mu)}$ follows

$$q_{cr} = \frac{q_{cr}^0}{1+\left[2q_{cr}^0(1+\mu)/E\right]}.$$

It means that the effects of transverse shear do not significantly affect the accuracy of calculations for a thin layer. In comparison with the study of isotropic plates, taking into account the influence of shear deformations becomes crucial in studies of anisotropic plates.

With taking into account the transverse shear, the deformations of a round plate with a loss of stability are described by the functions

$$\vartheta_r(r) = \vartheta_r^0(r), \; w' = \left(1+q_{cr}^0/B\right)\cdot\vartheta_r^0(r).$$

We consider the simplest case of anisotropy of the elastic properties of the material of the exfoliated part of the plates [4]. All planes parallel to the median plane of the plate are isotropy planes. We consider $|\sigma_z| << |\sigma_r|, |\sigma_\theta|$. We study the nonlinear behavior of circular delamination in a layered anisotropic plate.

The change in the total potential energy has the form

$$\Delta E = U_2 + V + \Pi_2 + T + U_4.$$

The inclusion of transverse shifts does not affect the formulas for calculating the quadratic deformations of the middle surface (U_4). T is the potential energy of shear strain.

From the stationary condition of the total potential energy follows

$$\eta\frac{q_{cr}^0}{\left(1+q_{cr}^0/B\right)^2} + \eta^3\frac{3}{14}\frac{Eh}{R^2\left(1+q_{cr}^0/B\right)^2} + \eta\frac{\left(q_{cr}^0/B\right)^2 B}{\left(1+q_{cr}^0/B\right)^2} - \eta\cdot q = 0.$$

We introduce dimensionless quantities

$$\bar{q} = q/q_{cr}^0; \; \bar{E} = \frac{E_1}{G_{12}}; \; \bar{h} = \frac{h}{R}; \; \bar{\eta} = \frac{\eta}{h}.$$

It can be seen from (3) that taking into account transverse shifts in the exfoliated part gives an order correction q_{cr}^0/B compared to unit. Determining the value q_{cr}^0, we evaluate this correction (3)

$$\frac{q_{cr}^0}{B} = \frac{14{,}68E_1h^3}{12\left(1-\mu_{12}\right)}\frac{1}{E_1h} \approx \frac{E_1}{G_{12}}\frac{h^2}{R^2}; \; c = 1+\frac{q_{cr}^0}{B} = 1+\bar{E}\bar{h}^2.$$

Table 1 shows the values of the order correction q_{cr}^0/B for various materials with different plate thicknesses and the same radius. Given $\eta \neq 0$, we come to the dependence

$$\bar{\eta} = \sqrt{\frac{56(1+\bar{E}\bar{h}^2)}{9(1-\mu_{12}\mu_{21})}\left\{\bar{q} - \frac{4-3\bar{E}\bar{h}^2(1-\mu_{12}\mu_{21}{}^2)}{4(1+\bar{E}\bar{h}^2)^2}\right\}}.$$

Table 2 shows the constant values for the materials.

Figure 1 shows the results of a study of the supercritical behavior of a defect of the type delamination from anisotropic materials.

Table 1. The values of the order correction q_{cr}^{0}/B for various materials.

Material	The order correction q_{cr}^{0}/B		
	$\bar{h}=0.01$	$\bar{h}=0.02$	$\bar{h}=0.03$
Graphite plastic	0.0067	0.027	0.0604
Kevlar-reinforced plastic	0.0028	0.0112	0.0252
Graphiteepoxide	0.00205	0.0082	0.0184
Fiberglass	0.0006	0.0025	0.0055

Table 2. The constant values for various materials.

Material	E_1 $(\mathrm{kN\,mm^{-2}})$	E_2 $(\mathrm{kN\,mm^{-2}})$	G_{12} $(\mathrm{kN\,mm^{-2}})$	μ_{12}
Graphite plastic	215	6.5	3.2	0.26
Kevlar-reinforced plastic	70	4.5	2.5	0.35
Fiberglass	53	14.0	8.6	0.26

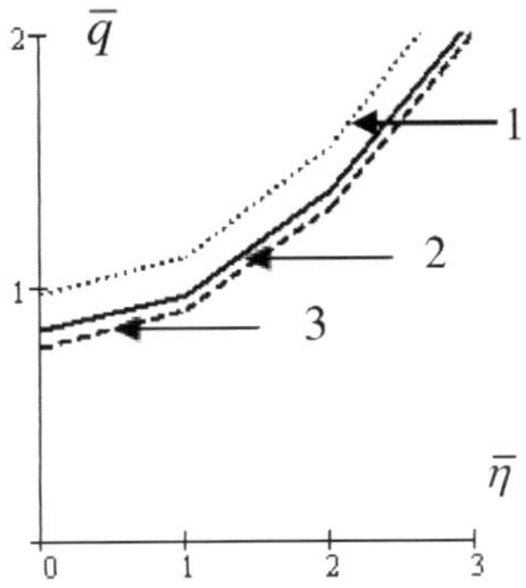

Figure 1. Numerical examples are obtained for: graphitoplastics (curve 3); plastic reinforced by Kevlar (curve 2); isotropic material (curve 1).

3. Analysis of the results of the calculation of a round plate

The inclusion of shear deformations in the study of the supercritical behavior of a round plate for isotropic materials is not practical, but it makes sense with extremely strong anisotropy of the elastic properties of the material. The effect of transverse shear effects leads to a decrease in the critical load. The experimental results are presented in [5-8].

Acknowledgments

The reported study was funded by the Russian Foundation for Basic Research (RFBR), project No. 18-29-18050/18 mk.

References

[1] Alfutov N A, Zinoviev P A and Popov B G 1984 *Calculation of Multilayer Plates and Shells of Composite Materials* (Moscow: Mashinostroenie)

[2] Ambartsumyan S A 1974 *General Theory of Anisotropic Shells* (Moscow: Science)

[3] Pobedry B E 1984 *Mechanics of Composite Materials* (Moscow: Moscow State University Press)

[4] Alfutov N A 1991 *Fundamentals of Calculation on the Stability of Elastic Systems* (Moscow: Mashinostroenie)

[5] Chermoshentseva A S, Bokhoeva L A and Rogov V E 2016 Experimental research of models of thin-walled-compressed rings with defects type of delamination of layered composite

IOP Conf. Series: Materials Science and Engineering **684** (2019) 012024 doi:10.1088/1757-899X/684/1/012024

materials *The 11th International Forum on Strategic Technology IFOST-2016 (Novosibirsk)* vol 1 (Red Hook, New York: Curran Associates, Inc) pp 112–4

[6] Bokhoeva L A, Rogov V E, Chermoshentseva A S and Lobanov D V 2016 Stability and process of destruction of compressed plate of layered composite materials with defects *Innovative Technologies in Engineering* **142** 12077

[7] Chermoshentseva A S, Pokrovskiy A M and Bokhoeva L A 2016 The behavior of delaminations in composite materials – experimental results *IOP Conf. Ser.: Mat. Sci. Eng.* **116** 012005

[8] Chermoshentseva A S, Pokrovskiy A M, Bokhoeva L A, Baldanov A B and Rogov V E 2019 Influence of modification by nanodispersed powders on layered composite aerospace hulls and protective shields *IOP Conf. Ser.: Mat. Sci. and Eng.* **560** 012178

IOP Conf. Series: Materials Science and Engineering **684** (2019) 012025 doi:10.1088/1757-899X/684/1/012025

On the results of integrated researches of the initial raw material and fenced glass structure with the application of modern physical and chemical methods

N N Anchiloev[1,3] **and V E Rogov**[2,4]

[1] East Siberia State University of Technology and Management, 40v Klyuchevskaya Street, Ulan-Ude 670013, Russia
[2] Baikal Institute of Nature Management, Siberian Branch of Russian Academy of Sciences, 6 Sakhyanova Street, Ulan-Ude 670047, Russia

e-mail: [3]anchiloev@yandex.ru, [4]rogov54v@mail.ru

Abstract. Studies have been conducted on the structures of the initial raw material and obtained on their basis foam glass of the "aluminosilicate rock –cullet – sodium hydroxide" systems with the use of X-ray and thermal analysis. The interrelation of the prescription-technological parameters and the physical and mechanical properties of the foam glass of the system "cullet – aluminosilicate rock – sodium hydroxide" has been established. Compositions of foam-forming mixtures have been developed, thus making it possible to obtain foam glass with the density from 412 to 1200 kg m^{-3} and strength from 2.0 to 16.1 MPa.

Depending on their functional purpose, all materials used in construction have various physical and mechanical properties. Any properties of the material are influenced by their chemical and phase compositions, production methods, as well as the internal structure [1-5]. Of all the above characteristics, the structure of the thermal insulation materials is crucial. This circumstance requiures a thorough study of the structure of the obtained materials and their starting components.

To study the structure, compositions and properties of the materials obtained, as well as the feedstock in a comprehensive way, the following physical and chemical methods were used: X-ray phase analysis; IR spectroscopy electron microscopy; thermal analysis. All studies were carried out on the basis of the Progress Collective Use Center of the Innovation Center of the FSBEI HE ESSUTM, as well as the Buryat Scientific Center of the RAS SB.

During the research and experimental design, clay and perlite deposits of the Buryat Republic, bar glass and sodium hydroxide were considered as raw materials.

X-ray phase analysis of cullet, as a rule, reveals their X-ray amorphous. The infrared spectra of cullet (figure 1) show the presence of a wide absorption band in the high-frequency region of the spectrum at 1050 cm^{-1}, which corresponds to the vibrations of relatively highly polymerized groups of $[SiO_4]$ tetrahedra. At a frequency of 772 cm^{-1} absorption bands indicate an increased degree of polymerization of silicon – oxygen groups upon incorporation of $[AlO_4]$ groups into the silicon and oxygen network. At 600 cm^{-1} the absorption band indicates the insignificant presence of various modifications of Si_2O~ in the cullet structure, which can be attributed to Si–O–Si bonds as the second tone of stretching vibrations. The presence of effects in the range of 1510–1679 cm^{-1} can be explained by the deformation

IOP Conf. Series: Materials Science and Engineering **684** (2019) 012025 doi:10.1088/1757-899X/684/1/012025

vibrations of free water molecules, adsorbed by glass. The absorption band at 2878 cm^{-1} may be caused by stretching vibrations of hydroxyl groups.

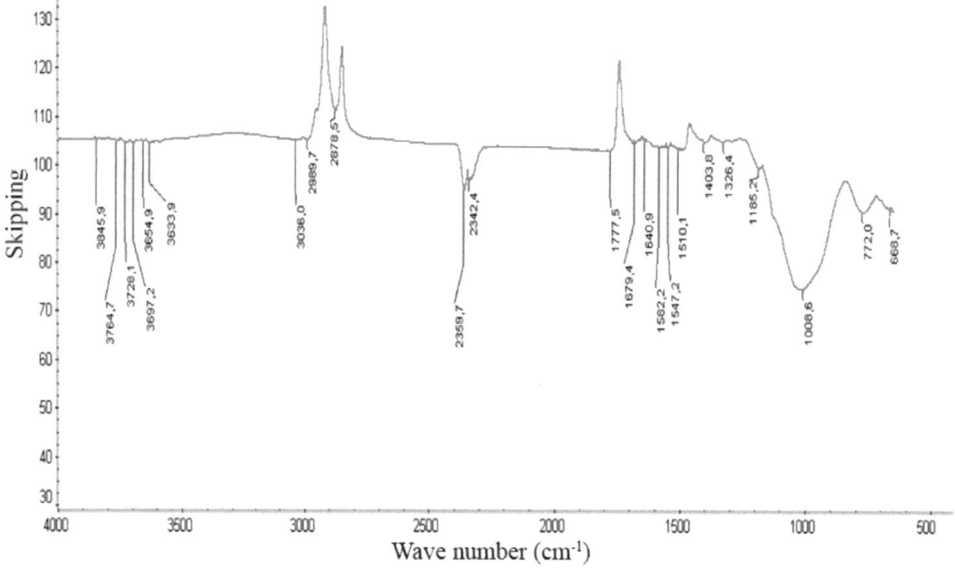

Figure 1. IR spectroscopy of cullet.

According to the electron microscopic studies, the structure of cullet is homogeneous with sufficient uniformity of its chemical composition.

According to X-ray powder diffraction data (figure 2), the structure of clay rocks is represented by montmorillonite (0.457; 0.447; 0.245; 0.256; 0.223; 0.167 nm), kaolinite (0.713; 0.446; 0.357; 0.259; 0.233 nm), calcite group minerals (1.424; 0.302; 0.289; 0.279; 0.213; 0.182; 0.159 nm), low-temperature quartz (0.425; 0.334; 0.228; 0.182; 0.154 nm), hydromica (0.996; 0.446; 0.285; 0.2558; 0.197; 0.178 nm), feldspars (0.3236; 0.318; 0.259; 0.178 nm).

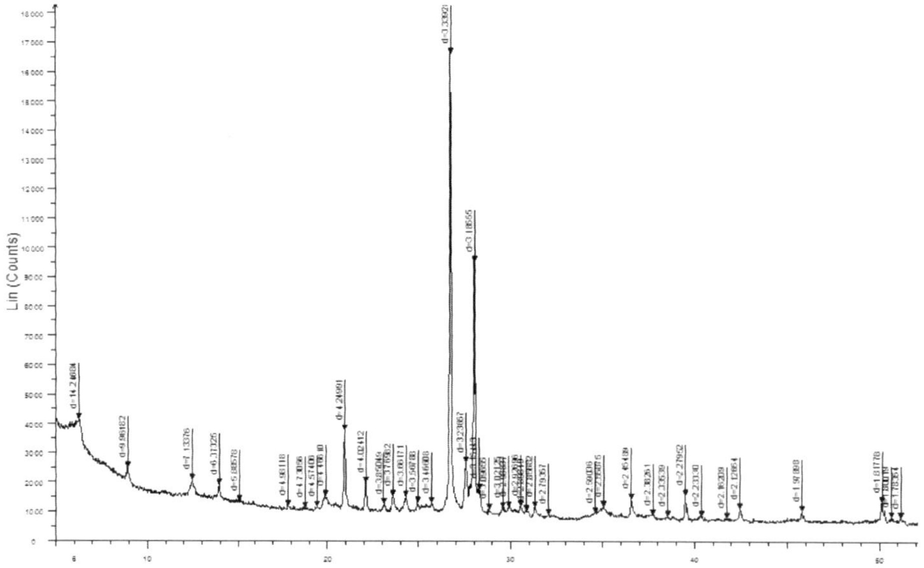

Figure 2. XRD clay.

IR spectroscopy of clay rocks (figure 3) shows that the presence of the absorption band at frequencies of 1000–1008 cm^{-1} in the structure of the starting materials is characterized by stretching vibrations of siloxane groups, and the presence of bands in the range 500–700 cm^{-1} is associated with the presence of silica modifications. The bands in the range of 3500-3600 relate to OH-stretching and deformation vibrations of free and bound water in the form of free hydroxyl.

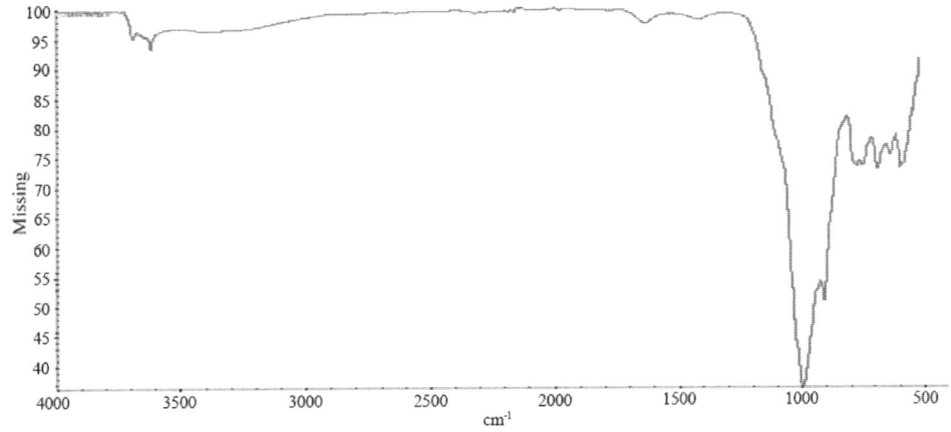

Figure 3. ICA – clay rock spectrum.

In the range from 4 to 40 $\theta°$ the roentgenogram of vitreous perlite (figure 4) demonstrates the peaks of low intensity, from which reflexes with d/n = 4.191; 3.346; 3.223; 2.817 Å can be distinguished. Hence it follows that glassy perlite is practically X-ray amorphous.

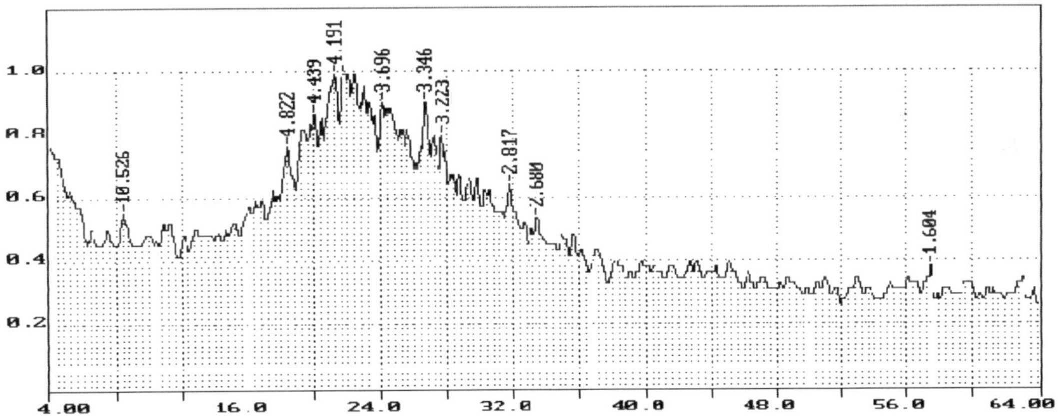

Figure 4. X-ray analysis of perlite rock.

Glass-perlite infrared spectroscopy (figure 5) showed the presence of a wide absorption band in the high-frequency region of the spectrum at 3500 cm^{-1}. The presence of weak doubles at frequencies of 2850 and 2910 cm^{-1} characterizes hydrogen bonds formed by hydroxyls bound to silicon atoms. The absorption band with a maximum at 1650 cm^{-1} corresponds to the deformation vibrations of O–H bonds and indicates the presence of molecular water and water in the form of free hydroxyl. An intense absorption band at 1050 cm^{-1} in pearlite rock indicates a strengthening of bonds in natural glass. Doubles at frequencies of 710–780 cm^{-1} are characteristic of absorption bands of crystalline quartz. The presence of various modifications of Si_2O~ in pearlite is also indicated by the absorption band with a maximum at the frequency of 475 cm^{-1}.

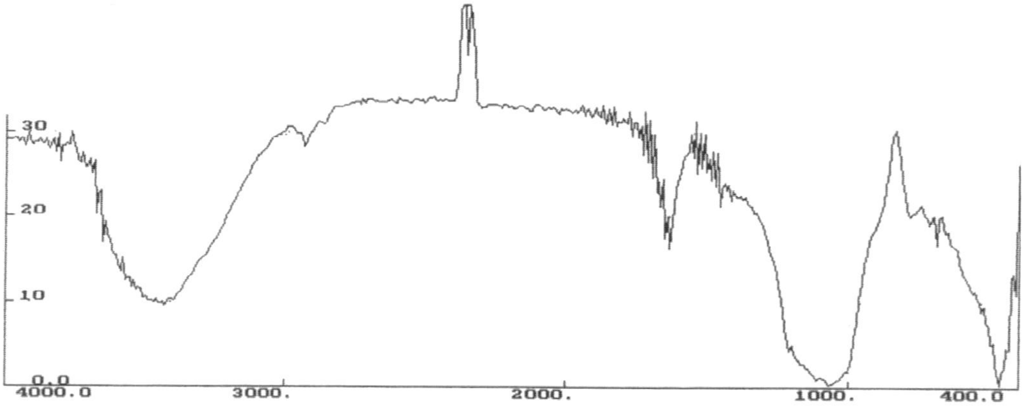

Figure 5. IR spectrum of perlite rock.

The materials studied were brought into a finely ground state by mechanical grinding in a VI 350x4 vibratory grinder. The glass charge for producing foam glass was prepared as follows: after preliminary crushing to the size of no more than 2 mm cullet pieces were dosed by weight in appropriate proportions and crushed in the vibratory grinder to the specific surface of 300–350 $m^2 kg^{-1}$.

After crushing clay rock and fine perlite were further crushed in the vibratory grinder to the specific surface of at least 350 and 450 $m^2 kg^{-1}$, respectively.

The components were closed with an aqueous solution of sodium hydroxide. The molding moisture content of the samples pressed was 16–19%. The prepared samples were subjected to heating and foaming in teh EKPS-10 laboratory muffle furnace with a maximum rise temperature of up to 1100 °C in accordance with the conventional heating, foaming and cooling modes.

It has been established that the introduction of aluminosilicate rocks — clay and perlite in the presence of an alkaline component into the composition of foaming mixtures during the production of foam glass, creates conditions for the directed regulation of the phase composition and the properties of the synthesized foam glass.

Along with the previously used methods, the results of the research prove the expediency of using mechanical and alkaline activation of thermal treatment of clay rock, which contributes to the intensified process of firing foam glass due to newly formed low-melting compounds. The use of X-ray phase analysis helps to study the structures of the starting raw materials, intermediate products, and finished foam glasses of the "aluminosilicate rock - cullet - sodium hydroxide" system.

In conclusion, we have developed the compositions of foaming mixtures that make it possible to obtain foam glass with an average density of 412–542 kg m^{-3} and compressive strength of 2.0–3.6 MPa for heat-insulating products and foam glass with an average density of 588–1200 kg m^{-3} and compressive strength of 1.7 –16.1 MPa for structural heat-insulating products at energetically favorable temperature conditions ranging from 850 °C to 900 °C.

Acknowledgments
The work was supported by the RFBR grant No. 18-29-18050/18 mk.

References
[1] Damdinova D R, Pavlov V E and Anchiloev N N 2014 *Building Materials* **8** 38–40
[2] Rogov V E, Kurbatov R V and Bokhoeva L A 2019 Thermochemical interaction of lead oxide with PTFE matrix *Journal of Applied Chemistry* **92(2)** 187–90
[3] Damdinova D R, Pavlov V E and Anchiloev N N 2017 *Bulletin of the Buryat State University. Chemistry, Physics* **2–3** 3–11
[4] Anchiloev N N 2016 *Mechanics of Structures and Materials: Proc. of 55th Scientific-Practical Conf. of Teachers, Researchers and Graduate Students of ESSUTM (Ulan-Ude)* issue 1

IOP Conf. Series: Materials Science and Engineering **684** (2019) 012025 doi:10.1088/1757-899X/684/1/012025

(Ulan-Ude: East Siberia State University of Technology and Management Press) pp 3–10
[5] Rogov V E, Tsyrenova G D, Perevalov A V and Bokhoeva L A 2016 Thermal interaction of lead oxide with PTFE *Chemical technology* **17(10)** 446–52

IOP Conf. Series: Materials Science and Engineering **684** (2019) 012026 doi:10.1088/1757-899X/684/1/012026

Strength evaluation of the relocatable building made of the polymer composite material

L A Bokhoeva, A B Baldanov and V Yu Kurokhtin

Strength of Materials Department, East Siberia State University of Technology and Management, building 1 40v Klyuchevskaya Street, Ulan-Ude 670013, Russia

e-mail: aldarbaldanov@gmail.com, kurokhtin91@gmail.com

Abstract. The article deals with the problem of improving the efficiency of the frame relocatable building structures by perfecting structural forms, introducing sound and lightweight materials, in particular, polymer composite materials (PCM). For this purpose, we have developed an engineering calculation algorithm using CAD / CAE systems and considered various options for deposition of PCM layers. The analytical formulas that determine the Eulerian instability for PCMs are compared with the values in ANSYS.

1. Introduction

A primary structural component of relocatable buildings and structures is a high-strength frame made of metal or wooden profiles. It is possible to increase further the efficiency of frame building structures through the improvement of structural forms and the use of strong and lightweight materials; currently, it is one of the urgent tasks of capital construction. The replacement of metal or wood profiles with multilayer polymer composite materials (PCM) is of particular interest for the construction industry. The use of this type of materials is especially relevant for construction of relocatable buildings in hard-to-reach places, as it is in oil and gas industry, geological exploration, where weight and frequent assembly and disassembly are of prime importance [1]. Such structures can include relocatable buildings.

However, despite good perspectives, PCMs have not found wide application in load bearing elements of building constructions. This is hampered by the complicated methods for calculating structures that do not find practical use. In this regard, it is relevant to create an engineering calculation that allows evaluating the behavior of a structure and its individual elements, as well as to develop computer methods for modelling and calculating building structures taking into account the PCM nonlinearity.

The paper considers the construction of a model of PCM-based relocatable building frame using modern CAD / CAE systems.

An analogue of this construction is the Mongolian yurt, with its frame made of wood. Lattice folding walls made of battens are pivotally fixed to each other, the yurt is covered with felt, over which a tarp is stretched. The roof is a dome formed of poles fixed in the center (figure 1(a)).

During the operation of this design, there is a loss of stiffness of the wooden frame and the stability of the elements, a decrease in strength during assembly-disassembly (2-3 times a year), the appearance of cracks at the joints of the rails and wear of the elements on the lower parts of the supports (figure

1(b)). To increase the strength and reduce the weight of the frame, wooden elements were replaced with PCM (figure 2).

(a) (b)

Figure 1. (a) traditional Mongolian tent construction "yurt"; (b) wear of wooden elements.

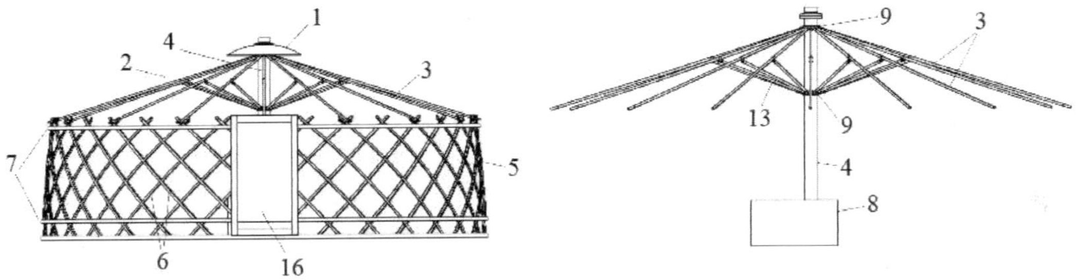

Figure 2. Relocatable building construction.

The improved construction contains a transparent dome (position 1 in figure 2), and roof (position 2), which has an umbrella type construction. The roof (position 2) frame consists of radially arranged rods (position 3), connected at one end to a center pillar (position 4), at the other end to a lathed wall (position 5) mounted in a circle and assembled from intersecting strips (position 6), which are fixed with belts (position 7). The center pillar (position 4) has a form of a hollow metal pipe (position 8). A patent for a utility model was obtained for the developed frame construction made from PCM [2].

2. Research methodology

To study the stress-strain state of the PCM structure, we used numerical methods of engineering analysis and application software packages Mathcad, SolidWorks, ANSYS, taking into account the characteristics of material and the actual loading conditions. A three-dimensional model of a tent-type construction using SolidWorks computer-aided design system (CAD) is developed. We carried out a series of numerical experiments taking into account the properties of the material, the direction of the fibers in each layer. The calculation procedure is presented in figure 3.

At the stage of specifying the preliminary model, we determined the form of the future structure and its elements related to the technology for manufacturing structural elements from PCM, operating conditions, research and development work. At the second stage, we calculated the elastic characteristics of PCM taking into account the filler and matrix material, the direction and size of fibers using Mathcad system. Next, we created a geometric model of the structure using the graphical interface of ANSYS program or CAD systems for complex objects (figure 4(a)). We carried out a preliminary check of the developed CAD-model of the construction, so that the structural elements made of PCM were of the "shell" type. At the last stage, we made a calculation of the construction.

Load summary was drawn up in accordance with the requirements of the Russian Set of Rules 20.13330.2016 [3] and presented in table 1. When calculating wind directions, snow loads, dead load of the structure were taken into account, but temperature climatic effects (small quantity of the 2nd degree) were disregarded.

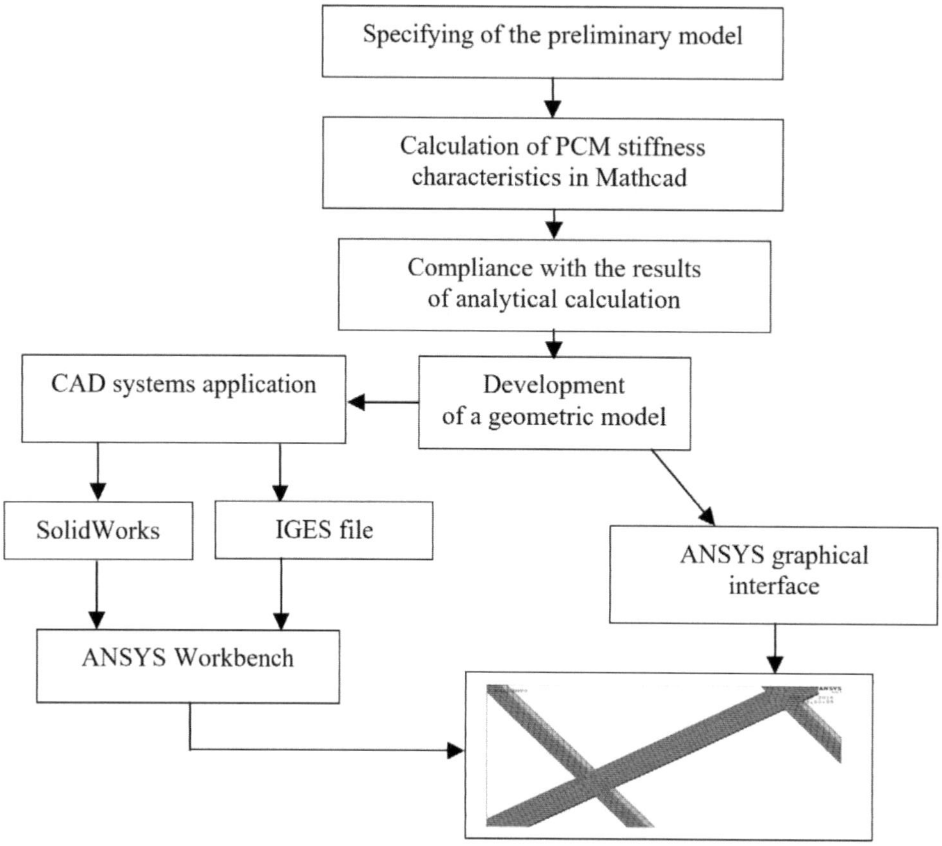

Figure 3. An algorithm for calculating the strength of structural elements made of PCM.

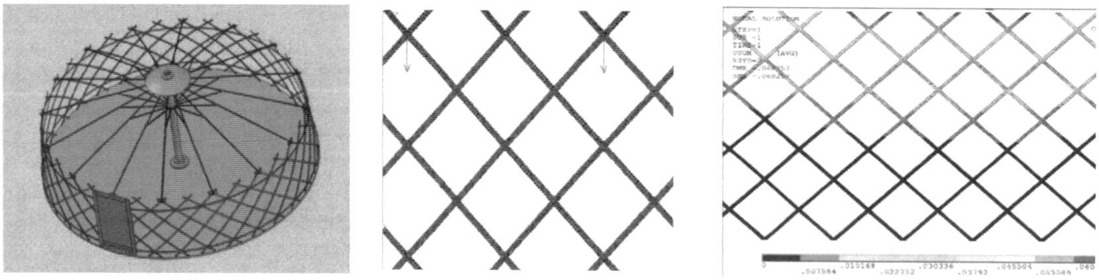

(a) (b) (c)

Figure 4. A strip made of the composite material: (a) 3D model of the relocatable building frame; (b) a finite element model; (c) equivalent von Mises stresses (Pa).

Table 1. Load calculation.

Area load per 1 m^2 (kPa)	Snow loads 1	Snow loads 2	Snow loads 3	Wind loads 1
	3.5	5.34	4	–0.545

As an example, we have selected a section of the frame – a lathed wall of the relocatable structure, which is under the maximum load. The frame of the structure is presented as a package of layers, with the properties of individual layers (monolayer is made of fiberglass) specified (figure 5). Fiberglass has the following characteristics: $E_1 = 5.4 \cdot 10^4$ MPa, $E_2 = 1.2 \cdot 10^4$ MPa, $G_{12} = 0.5 \cdot 10^4$ MPa, $\mu = 0.28$. The monolayer thickness is b = 0.2 mm. The scheme of reinforcement is $[0_{12}/90_8/\pm45_{10}]$.

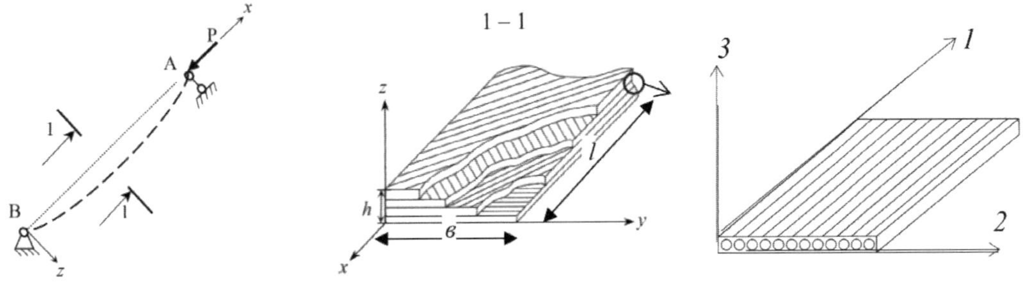

Figure 5. A strip made of composite material: (a) design pattern of a rod made of composite material; (b) cross section of a rod made of composite material; (c) monolayer.

We have determined the stiffness characteristics of the composite package by the known stiffness characteristics of the monolayer [4]:

$$E^k{}_{11} = \bar{E}_1 c^4 + \bar{E}_2 s^4 + G_{12} s_2^2 + \mu_{21} \bar{E}_1 2s^2 c^2;\; E^k{}_{22} = \bar{E}_1 s^4 + \bar{E}_2 c^4 + G_{12} s_2^2 + \mu_{21} \bar{E}_1 2s^2 c^2;$$

$$E^k{}_{12} = \left(\bar{E}_1 + \bar{E}_2\right) s^2 c^2 - G_{12} s_2^2 + \mu_{21} \bar{E}_1 \left(c^2 + s^2\right);\; E^k{}_{33} = \left(\bar{E}_1 + \bar{E}_2\right) s^2 c^2 + G_{12} s_2^2 - \mu_{21} \bar{E}_1 2s^2 c^2;$$

$$E^k{}_{13} = \bar{E}_1 s c^3 - \bar{E}_2 s^3 c - G_{12} s_2 c_2 + \mu_{21} \bar{E}_1 \left(s^3 c - s c^3\right);$$

$$E^k{}_{23} = \bar{E}_1 s^3 c - \bar{E}_2 s c^3 + G_{12} s_2 c_2 + \mu_{21} \bar{E}_1 \left(s c^3 - s^3 c\right),$$

where $c = \cos\phi$; $c_2 = \cos 2\phi$; $s = \sin\phi$; $s_2 = \sin 2\phi$; $\bar{E}_1 = E_1 / (1 - \mu_{12}\mu_{21})$; $\bar{E}_2 = E_2 / (1 - \mu_{12}\mu_{21})$; $E_1\mu_{21} = E_2\mu_{12}$.

Here $E_1, E_2, G_{12}, \mu_{12}$ are the stiffness characteristics of the monolayer. The approximate stiffness characteristics of the multi-layered CM with a sufficient accuracy for design calculations are determined by the formulas:

$$E_x = A_{11} - \frac{A_{12}^2}{A_{22}};\; E_y = A_{22} - \frac{A_{12}^2}{A_{11}};\; G_{xy} = A_{33};\; \mu_{xy} = \frac{A_{11}}{A_{22}};\; \mu_{yx} = \mu_{xy}\frac{E_y}{E_x} = \frac{A_{12}}{A_{11}},$$

where $A_{ij} = \frac{1}{H}\sum_{k=1}^{n} h_k E_{ij}^k$; $i, j = 1,2,3$; H – thickness of laminated package; h_k – thickness; k – layer; n – total number of layers in a package. The results of filling out these formulas in Mathcad system are presented in table 2.

Table 2. The stiffness characteristics of the multi-layered CM.

E_x (Pa)	E_y (Pa)	G_{xy} (Pa)	μ_{xy}	μ_{yx}
$3.924 \cdot 10^{10}$	$3.248 \cdot 10^{10}$	$7.118 \cdot 10^9$	0.248	0.205

Further, we calculated rods in ANSYS Mechanical system. To calculate the fiberglass rod, SOLID-46 multilayer element is used. This is a multilayer version of an 8-node three-dimensional solid element with three degrees of freedom for each node (U_x, U_y, U_z). The element is designed to simulate multilayer shells or multilayer solids [5]. The numerical solution of the nonlinear problem was carried out by the stepwise method. This method is based on the non-linear static calculation with a gradual increase in the load to the level at which the model becomes unstable. This allows to include such material features in the model as initial form deviations, plastic properties, gaps, etc.

The results of calculating the strain-stress state of the frame are presented in figure 4(c). The analysis of the numerical experiment shows that the upper part of lathed wall carries the highest load. Peak axially displacements in U_y-direction are observed in the setting of [0°/60°/120°] reinforcement scheme and are 0.5 mm modulo. The least U_y axial displacement takes place, when reinforcement scheme is [0°/±45°], and makes 0.467 mm modulo. The maximum stress on the supports of strips is 195.66 MPa. When reinforcing [0°/±45°], the structural frame can withstand maximum stresses of 82.989 MPa. The calculation results are presented in table 3.

Table 3. Maximum normal stresses and displacements in the lathed wall made of PCM at various reinforcement schemes.

Results of numerical calculations	[0°/90°/±45°]	[0°/60°/120°]	[0°/±45°]
Normal displacements U_x (mm)	0.253	0.251	0.252
Normal displacements U_y (mm)	–0.475	–0.499	–0.467
Normal displacements (mm)	0.097	0.097	0.097
Normal stresses σ_x (MPa)	–127.760	–125.690	–123.660
Normal stresses σ_y (MPa)	–85.400	–85.927	–82.989
Normal stresses σ_z (MPa)	–193.930	190.330	–195.660

The loss of stability is dangerous for thin-walled rods made of PCM, [6]; for their study we considered a linearized equation of quartic derivative in the following form:

$$v^{IV} + k^2 v'' = 0, \tag{1}$$

where $k^2 = P / E_x I_{min}$; $D = \dfrac{E_x h^3}{12\left(1-\mu_{xy}\mu_{yx}\right)}$; I_{min} – axial moment of inertia; E_x – elastic modulus; h – rod thickness; μ_{xy}, μ_{yx} – Poisson's ratio in the corresponding axis direction.

The solution to equation (1) is following:

$$v = A_1 \sin(kx) + A_2 \cos(kx) + A_3 x + A_4, \tag{2}$$

where A_1, A_2, A_3, A_4 are constants. Four boundary conditions are specified for this rod depending on the fixation of ends: 1) $v(0)=0$; 2) $v''(0)=0$; 3) $v(l)=0$; 4) $v''(l)=0$. Substituting the boundary conditions into the general solution (2), we obtain a system of four homogeneous linear equations (3) in the A_i unknowns.

$$\begin{cases} A_1 \sin(0) + A_2 \cos(0) + A_3 \cdot 0 + A_4 = 0; \\ -A_1 k^2 \sin(0) - A_2 k^2 \cos(0) = 0; \\ -A_1 \sin(kl) + A_2 \cos(kl) + A_3 l + A_4 = 0; \\ -A_1 k^2 \sin(kl) - A_2 k^2 \cos(kl) = 0. \end{cases} \tag{3}$$

If the determinant of this system (3) is equal to zero, we can find the eigenvalues of the P_n problem and the corresponding eigenfunctions. The smallest eigenvalue gives the critical load value, and the corresponding eigenfunction describes the form of the curved rod axis at loss of stability:

$$v_n(x) = A_1 \cdot \sin\left(\frac{\pi n}{l} \cdot x\right).$$

We have obtained the value of Euler's critical load for a rod at the loss of stability:

$$\frac{P \cdot 12\left(1 - \mu_{xy}\mu_{yx}\right)}{E_i h_i^3} = \left(\frac{\pi n}{l}\right)^2,$$

then for $n = 1$:

$$P_{cr} = \frac{\pi^2 E_x h_i^3}{12 \cdot l^2 \left(1 - \mu_{xy} \cdot \mu_{yx}\right)}.$$

The compared critical load values depending on the length of rods, which have been obtained analytically and in ANSYS, result in allowable convergence for engineering calculations (figure 6).

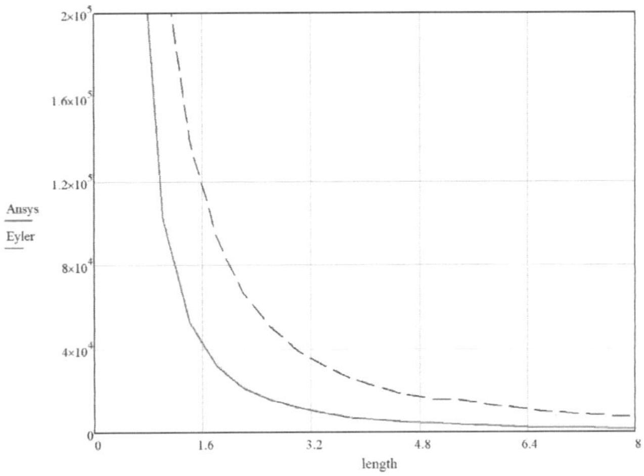

Figure 6. Euler's critical load for [0°; ±45°] reinforcement scheme and the critical load obtained in ANSYS.

Using this diagram, we can determine the permissible length of relocatable building rods for a given type of laying.

3. Conclusion

Numerical calculations confirmed our preliminary analytical calculations for the application of various materials and reinforcement schemes. For fiberglass quasi-isotropic reinforcement scheme is the best in terms of strength and stiffness of the product received. The reinforcement scheme [0°; ±45°] provides sufficient stiffness of the structure, maximum displacements of the frame strips does not exceed 0.467 mm, so that it corresponds to the strength of wooden structures. In addition, in the improved construction made of PCM 42 strips are used instead of 63 wooden, as a result its weight decreases by 2.5 times.

Acknowledgments
The work was supported by the state target program of the Ministry of Science and Higher Education of the Russian Federation, project No. 9.7667.2017/8.9, and by the grant "Young Scientists of ESSUTM – 2019".

References
[1] Bokhoeva L A and Baldanov A B 2015 Computer modelling of the easy figurative dwelling from composite materials taking into account climate of Baikal region *Issues in Modern Machines Mechanics: Proc. of the VI Int. Conf. (Ulan-Ude)* vol 3, ed V S Balbarov (Ulan-Ude: East Siberia State University of Technology and Management Press) pp 116–21
[2] Baldanov A B and Bokhoeva L A 2012 *The Yurta Framework – the Collapsible Dwelling: Russia patent No. 123429* (Moscow: Russian Federal Service for Intellectual Property)
[3] 2016 *Set of Rules SP 20.13330.2016. Loads and Actions* (Moscow: Standartinform)
[4] Bokhoeva L A 2007 *Features of the Calculation of the Strength of Structural Elements of Isotropic and Composite Materials with Valid Defect* (Ulan-Ude: East Siberia State Technological University Press)
[5] Basov K A 2002 *ANSYS in Examples and Problems* ed D G Kraskovskii (Moscow: KompiyuterPress)
[6] Chermoshentseva A S, Pokrovskiy A M and Bokhoeva L A 2016 *IOP Conf. Ser.: Mater. Sci. Eng.* **116** 012005

IOP Conf. Series: Materials Science and Engineering **684** (2019) 012027 doi:10.1088/1757-899X/684/1/012027

An experimental research of the laminated composite plates stability

L A Bokhoeva and E B Bochektueva

Strength of Materials Department, East Siberia State University of Technology and Management, 40v building 1 Klyuchevskaya Street, Ulan-Ude 670013, Russia

e-mail: bohoeva@yandex.ru, bochektueva.e@yandex.ru

Abstract. The article shows the results of experimental studies of stability, supercritical behaviour and fracture of plates of a layered composite material. Composite plates are made of fiberglass layers glued with epoxy resin. To reliably assess the bearing capacity of laminated plates, numerical and experimental methods are used. A methodology for experimental research of the deformation and strength properties of composite materials using testing equipment has been developed. The difference between the critical strength values of three-layer samples and the critical strength of four-layer samples was determined to be 2-3%. Thus, products from composite materials (unmanned aerial vehicles) can be made from three-layer plates to reduce their weight. Models of two-layer, three-layer, and four-layer plates were performed with the ANSYS program. The obtained experimental results were compared with numerical values. The possibility of using experimental data of the mechanical behaviour of layered composite materials in the design and calculation of unmanned aerial vehicles was estimated.

1. Introduction

One of the main tasks facing designers to create unmanned aerial vehicles is to increase the reliability of the structure [1-2] due to the stability of strength characteristics and the accuracy of aerodynamic contours with a simultaneous increase in manufacturability, as well as creating a structure with a more accurate profile from composite materials [3-4]. Preliminary calculations, experimentally confirmed simulation, introduction of nanomodifications of composite materials allow us to develop a new approach to extending the life cycle of elements and assemblies, increasing the specific strength, rigidity and stability of the structure when working under complex loading (including dynamic) [5], and above all, these lead to a decrease in the mass of structures with an increase in their bearing capacity. Calculations of the stress-strain state of structural elements and the optimal choice of the layering structure of composite materials ensure the strength of the product depending on the perception of external loads. In selecting rational structural and technological solutions for the products, modeling the stress-strain state, optimization and experimental characterization of layered composite materials are of great importance [6]. Layered composite materials have a wide range and a unique combination of properties such as high strength, corrosion resistance, wear resistance [7-8]. A significant role in the process of creating new materials is played by an experimental study on the strength and stability of samples [9].

2. Experimental work

This article presents the results of experimental studies of the stability and fracture of laminated plates made of composite material with a different number of layers. Two series of samples were made, one from fiberglass and the other with the inclusion of foam between the layers of fiberglass. Samples of the first series of fiberglass are:

- 10 samples of four layers;
- 10 samples of three layers;
- 10 samples of two layers.

Samples were made in the research and production laboratory "Reliability, strength of products and structures" of the East Siberia State University of Technology and Management. The manufacturing of samples involved contact molding by layering in a mold, lubricated with a release coating of reinforcing material (T-10 fiberglass) with simultaneous impregnation of each layer with a binder (epoxy resin with hardener). After impregnation additional rolling was carried out to eliminate air bubbles and ensure the uniform distribution of the resin. At this point a thorough air removal is necessary to avoid interlayer defects that otherwise may occur. Batches of samples consisting of two, three and four layers were made. After forming they were dried and subjected to mechanical processing. Then all the samples were numbered and measured at the cross section of their working part. Using, The width and thickness in different sections of the sample were measured by a micrometer and an electronic caliper. Deviations both in width and thickness were ±0.1 mm.

Samples for the experimental determination of the mechanical characteristics of composite materials satisfy the following requirements [9]:

- the ability to implement simple standard loading schemes;
- simplicity and low cost of the device for testing;
- ease of installation in the testing machine and testing;
- insensitivity to the method of attachment;
- reproducibility of experimental estimates of the studied characteristics.

Figure 1 shows the manufacture and preparation of samples for testing.

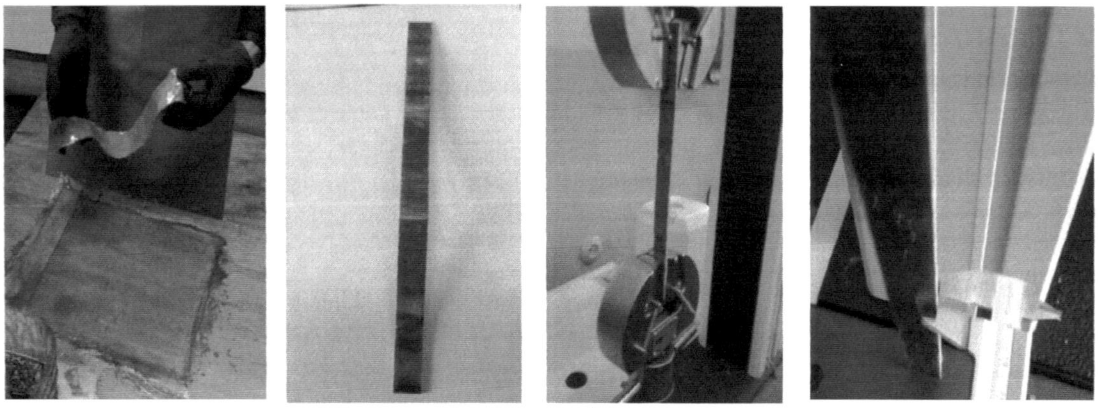

Figure 1. Production and preparation of samples for experiments.

Accurate determination of the mechanical characteristics of composite materials is possible using modern equipment with high accuracy of recording loads and movements. The experimental work was carried out on the universal Instron 3367 machine. During compression tests, the test specimen is installed in the grips between the movable and fixed traverses. The loading of the sample is carried out by moving the active capture in a given mode. During the test it is possible to implement loading modes of the sample with a constant rate of change in stress, force, movement of the active capture or deformation. The measured parameters (forces and displacement) are converted by the sensors and the

IOP Conf. Series: Materials Science and Engineering **684** (2019) 012027 doi:10.1088/1757-899X/684/1/012027

control unit of the machine into electrical signals, which are displayed on the monitor screen of the control computer in the corresponding tension or compression diagrams. Figure 2 shows photographs of test equipment for compressing plates.

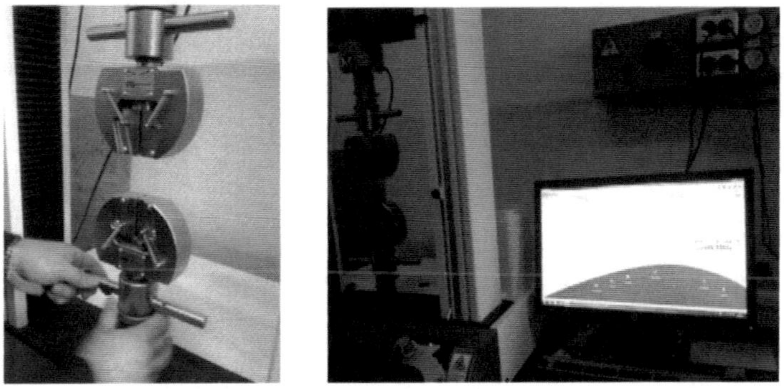

Figure 2. Test machine Instron 3367.

3. The results of experimental work

The first batch of samples consisted of ten four-layer plates of the following sizes: t = 2 mm, h = 19.6 mm, l = 170 mm. The plate was installed in the clamps and the load was set at the speed of v = 50 mm min^{-1}. When the critical force was reached (figure 3), the plate lost stability but continued to take the load until it collapsed. The critical load, the load during the destruction of the sample was determined. The results of four-layer samples are presented in figure 4.

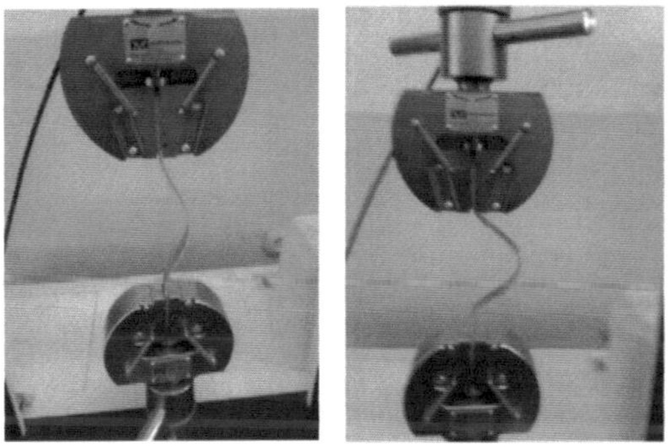

Figure 3. Compression test of two-layer plates.

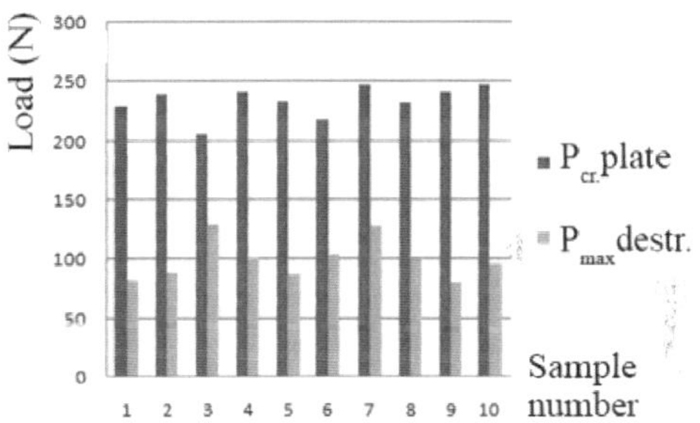

Figure 4. Compression of four-layer samples numbered 1 to 10.

The second batch consisted of ten three-layer plates (samples No. 11 to 20) of the following sizes: t = 2.5 mm, h = 20 mm, l = 170 mm. The results are presented in figures 5, 6.

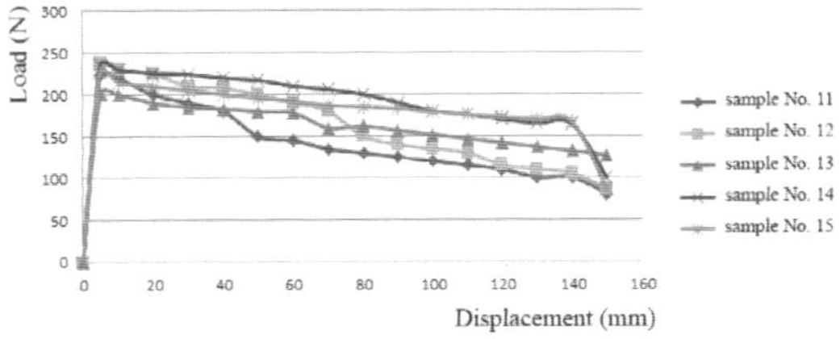

Figure 5. Compression of three-layer samples numbered 11 to 15.

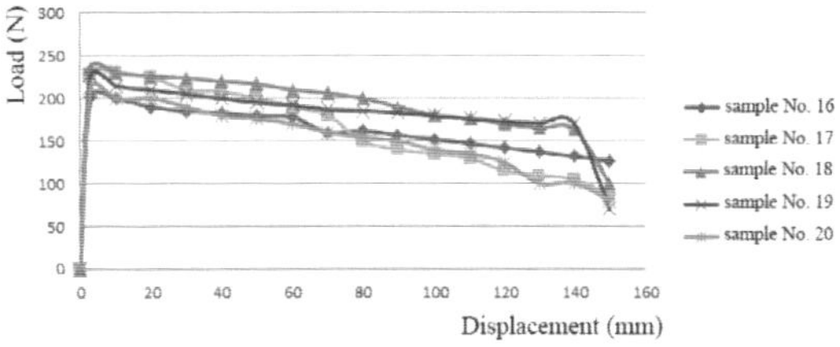

Figure 6. Compression of three-layer samples numbered 16 to 20.

IOP Conf. Series: Materials Science and Engineering **684** (2019) 012027 doi:10.1088/1757-899X/684/1/012027

The third batch consisted of ten two-layer plates (t = 1 mm, h = 20 mm, l = 170 mm, v = 50 mm min^{-1}), samples No. 26 to 35 are presented in figures 7, 8.

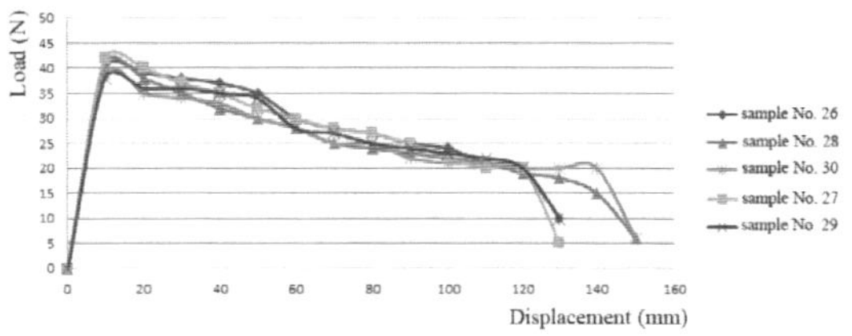

Figure 7. Compression of two-layer samples numbered 26 to 29.

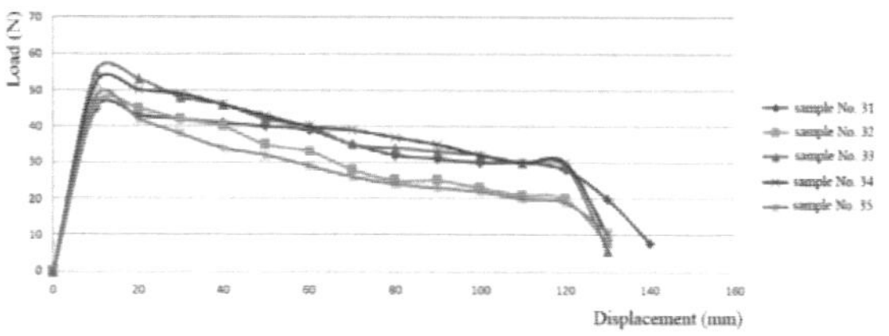

Figure 8. Compression of two-layer samples numbered 31 to 35.

The experimental results are presented in table 1. The critical load of the sample is designated as "$P_{cr.}$ plate". The load upon destruction of the sample is designated as "$P_{cr.}$ destr".

Table 1. Sample test results.

	Four-layer samples									
Sample No.	1	2	3	4	5	6	7	8	9	10
$P_{cr.}$ plate (N)	230	240	206	241	234	219	248	232	241	248
$P_{cr.}$ destr. (N)	82	89	130	102	87	105	128	100	81	97
	Three-layer samples									
Sample No.	11	12	13	14	15	16	17	18	19	20
$P_{cr.}$ plate (N)	223	238	202	236	227	213	216	225	236	241
$P_{cr.}$ destr. (N)	80	86	127	100	86	102	125	98	78	94
	Two-layer samples									
Sample No.	26	27	28	29	30	31	32	33	34	35
$P_{cr.}$ plate (N)	40	42	41	38	39	45	46	55	52	48
$P_{cr.}$ destr. (N)	10	5	6	11	6	8	8	6	10	11

4. Computer simulation of layered composite plates

Two-layer, three-layer and four-layer plates were simulated by ANSYS. The results of numerical calculation by the finite element method are presented in figures 9 and 10 for the deformed state and stress state of a three-layer plate, respectively.

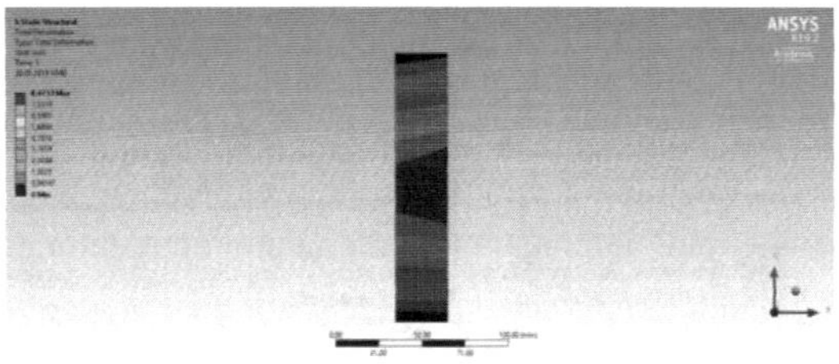

Figure 9. Deformed state of a three-layer plate in the ANSYS system (mm).

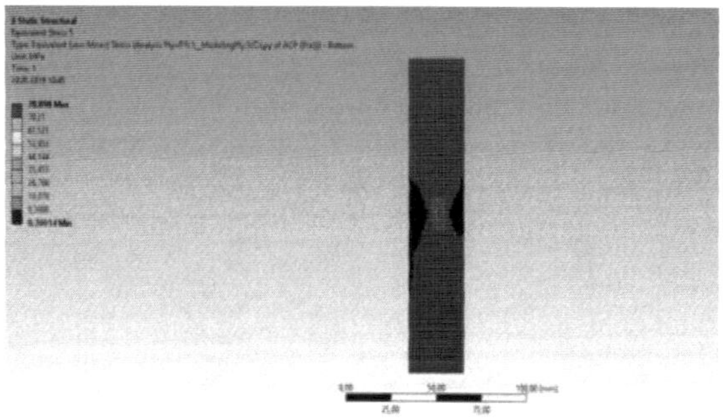

Figure 10. The stress state of a three-layer plate in the ANSYS system (MPa).

Table 2 presents the results of experimental values of the critical load of the samples and numerical values obtained by the finite element method in the ANSYS package. The results obtained showed good convergence between analytical and numerical values.

Table 2. The results of experimental and numerical values of the critical load and fracture load.

Number of layers	$P_{cr.}$ (N) (experimental value)	$P_{cr.}$ (N) (numerical value)
2	45	48.6
3	223	239

5. Findings

A set of experiments was performed with samples from multilayer composite materials and the following results were obtained:

- Four-, three- and two - layer samples from composite materials were tested. The values of the critical load and that of the load at failure were obtained.
- A manufacturing procedure of samples was developed. It was found that the discrepancy between the critical load of four-layer samples and that of three-layer samples was 2–3%. Therefore, three-layer plates can be used to reduce the weight of the products made from composite materials. The discrepancy between the critical load of three-layer samples and the critical load of two-layer samples was 19–21%. The discrepancy between the experimental and numerical values was 7–8%.

Acknowledgments
The work was supported by the grant "Young Scientists of the ESSUTM – 2019" and by the state task of the Ministry of Science and Higher Education of Russia, project No. 9.11221.2018/11.12.

References

[1] Andreev Yu V et al. 2015 Problems of standardization of the strength of UAV structures *Proc. of Scientific and Technical Conf. "Strength of Aircraft Structures" (Proc. of TsAGI)* (Zhukovsky: Central Aerohydrodynamic Institute Press) pp 69–78
[2] Bolsunovsky A L, Buzover N P, Kazhan V G, Kazhan A V and Fomin V M 2005 Calculated and experimental studies of the aerodynamic characteristics of the thematic UAV model *Materials of the XVI School-seminar "Aerodynamics of Aircraft"* (Zhukovsky: Central Aerohydrodynamic Institute Press) p 24
[3] Villasenor J 2014 "Drones" and the future of domestic aviation *Proceedings of the IEEE* **102(3)** 235–8
[4] Gupta L, Jain R and Vaszkun G 2016 Survey of important issues in UAV communication networks *IEEE Communications Surveys & Tutorials* **18(2)** 1123–52
[5] Bokhoeva L A, Bochektueva E B, Zaitsev V A, Perevalov A V and Chermoshentseva A S 2012 Dynamic testing of aircraft products *Materials of the V Int. Conf. "Issues in Modern Machines Mechanics"* vol 2 (Ulan-Ude: East Siberia State University of Technology and Management Press) pp 65–9
[6] Bokhoeva L A and Bochektueva E B 2018 Development of a multi-channel measuring system for experimental research of aeronautical products *Materials of the VII Int. Conf. "Issues in Modern Machines Mechanics"* vol 1 (Ulan-Ude: East Siberia State University of Technology and Management Press) pp 151–5
[7] Sarbaev B S and Krivolutskaya I I 2015 A method for calculating ultimate stresses for multilayer fibrous composites in the plane stress state *Construcrions of Composite Materials* **2** 3–9
[8] Pokrovsky A M and Chermoshentseva A S 2017 An experimental study of the effect of nano-

additives on the properties of composite materials with interlayer defects *Aerospace MAI Journal* **24(3)** 212–21
[9] Bokhoeva L A and Damdinov T A 2006 Determination of critical loads by energy method with shear deformations *Vestnik KGTU im. A.N. Tupoleva* **1** 3–7

AUTHOR INDEX